MICROBIAL INOCULATION
OF
CROP PLANTS

Special Publications of the Society for General Microbiology

Publications Officer: Dr Duncan E.S.Stewart-Tull, Harvest House, 62 London Road, Reading RG1 5AS, UK

Publisher: Academic Press

1. Coryneform Bacteria
2. Adhesion of Micro-organisms to Surfaces
3. Microbial Polysaccharides and Polysaccharases
4. The Aerobic Endospore-forming Bacteria: Classification and Identification
5. Mixed Culture Fermentations
6. Bioactive Microbial Products: Search and Discovery
7. Sediment Microbiology
8. Sourcebook of Experiments for the Teaching of Microbiology
9. Microbial Diseases of Fish
10. Bioactive Microbial Products 2: Development and Production
11. Aspects of Microbial Metabolism and Ecology
12. Vectors in Virus Biology
13. The Virulence of Escherichia coli
14. Microbial Gas Metabolism
15. Computer-Assisted Bacterial Systematics
16. Bacteria in Their Natural Environments
17. Microbes in Extreme Environments
18. Bioactive Microbial Products 3: Downstream Processing

Publisher: IRL Press

19. Antigenic Variation in Infectious Diseases
20. Nitrification
21. Carbon Substrates in Biotechnology
22. Gene Structure in Eukaryotic Microbes
23. Spatial Organization in Eukaryotic Microbes
24. Bacterial Infections of Respiratory and Gastrointestinal Mucosae
25. Microbial Inoculation of Crop Plants

This book is based on a symposium of the SGM held in April 1988.

SPECIAL PUBLICATIONS OF THE SOCIETY FOR GENERAL MICROBIOLOGY
—————————— VOLUME 25 ——————————

MICROBIAL INOCULATION OF CROP PLANTS

Edited by

R.Campbell

Department of Botany, University of Bristol, Bristol BS8 1UG, UK

R.M.Macdonald

AFRC Institute of Arable Crop Research, Rothamsted Experimental Station, Harpenden, Herts AL5 2JQ, UK

1989
Published for the
Society for General Microbiology
by

OXFORD UNIVERSITY PRESS
Oxford New York Tokyo

IRL Press at Oxford University Press
Eynsham
Oxford OX8 1JJ
England

©1989 Society for General Microbiology

First Published 1989

All rights reserved by the publisher. No part of this book may be reproduced or transmitted in any form by any means, electronic or mechanical, including photocopying, recording or any information storage and retrieval system, without permission in writing from the publisher.

British Library Cataloguing in Publication Data

Microbial inoculation of crop plants.
  1. Plants. Diseases. Biological control
  I. Campbell,R. II. Macdonald,R.M. III. Series 632'.96
  ISBN 0-19-963016-X
  ISBN 0-19-963017-8 pbk

Library of Congress Cataloging-in-Publication Data

Microbial inoculation of crop plants. / edited by R. Campbell and R.M. Macdonald.
  p. cm. — (Special publications of the Society for General Microbiology; v. 25)
  Bibliography: p.
  Includes index.
  ISBN 0-19-963016-X : $80.00. — ISBN 0-19-963017-8 (pbk.) : $50.00
  1. Microbial inoculants.   2. Crops—Inoculation.
  I. Campbell, R.E. (Richard Ewen) II. Macdonald, R.M.
  III. Series: Special publications of the Society for General Microbiology; 25.
  QR51.M44   1989
  630'.2'76—dc19

89-2067 CIP

Previously announced as:

ISBN 1-85221-201-2 (hardbound)
ISBN 1-85221-197-X (softbound)

Typeset and printed by Information Press Ltd, Oxford, England

Contents

Preface — vii

Contributors — ix

Abbreviations — xi

1. An overview of crop inoculation
   R.M. Macdonald — 1

2. An industrial view of microbial inoculants for crop plants
   G. Lethbridge — 11

3. Global importance of *Rhizobium* as an inoculant
   A.R.J. Eaglesham — 29

4. Present and future value of mycorrhizal inoculants
   D.P. Stribley — 49

5. The use of microbial inoculants in the biological control of plant diseases
   R. Campbell — 67

6. The use of plant viruses as inoculants
   R.F. White and J.F. Antoniw — 79

7. The use of blue-green algae and *Azolla* in rice culture
   B.A. Whitton and P.A. Roger — 89

8. The use of engineered and genetically distinct inocula on plants
   J.E. Beringer, K.A. Powell and N.J. Poole — 101

Index — 109

Preface

This Special Publication is the result of a meeting of the Ecology Group of the Society for General Microbiology at the University of Warwick, UK, on 14 April 1988. The aim was to review the rapidly developing subject of microbial inoculants of crop plants in the light of an expanding research effort worldwide, recently increased interest from commercial companies, and public concern over the release of microorganisms into the environment. The contributors were asked to briefly review the history of microbial inoculants, then to assess their present importance and to speculate on what might happen in the foreseeable future. It was hoped that this would enable research and development projects to be realistically assessed and directed into worthwhile areas.

The meeting was largely organized by Rod Macdonald who should receive such praise as is due for the content. Sadly, ill health forced Rod to retire during the course of publication, so editorial errors and omissions should be blamed on Richard Campbell.

We would like to thank the then convenor of the Ecology Group, Dr J.Gwyn Jones, and members of the Committee, for their help and support in arranging the meeting on which this special publication is based. We also thank Dr D.E.S.Stewart-Tull, Publications Officer of the Society for General Microbiology, for his help and advice in editing the book.

Finally, we thank all the contributors (especially those who handed in their typescripts on time!) for the high standard of their copy which reduced the labour of editing.

R.Campbell and R.M.Macdonald
Past Members of the Ecology Group Committee
Co-chairmen of the Meeting

Contributors

J.F.Antoniw
Plant Pathology Department, AFRC Institute of Arable Crop Research, Rothamsted Experimental Station, Harpenden, Herts AL5 2JQ, UK

J.E.Beringer
Microbiology Department, Medical School, University Walk, University of Bristol, Bristol BS8 1TD, UK

R.Campbell
Department of Botany, University of Bristol, Bristol BS8 1UG, UK

A.R.J.Eaglesham
Boyce Thompson Institute for Plant Research at Cornell University, Tower Road, Ithaca, NY 14853, USA

G.Lethbridge
Physiology and Biochemistry Division, Shell Research Limited, Sittingbourne Research Centre, Sittingbourne, Kent ME9 8AG, UK

R.M.Macdonald
AFRC Institute of Arable Crop Research, Rothamsted Experimental Station, Harpenden, Herts AL5 2JQ, UK

K.A.Powell
ICI Agrochemicals, Jealott's Hill Research Station, Jealott's Hill, Bracknell, Berks RG12 6EY, UK

N.J.Poole
ICI Agrochemicals, Jealott's Hill Research Station, Jealott's Hill, Bracknell, Berks RG12 6EY, UK

P.A.Roger
International Rice Research Institute, PO Box 933, Manila, The Philippines

D.P.Stribley
AFRC Institute of Arable Crop Research, Rothamsted Experimental Station, Harpenden, Herts AL5 2JQ, UK

R.F.White
Plant Pathology Department, AFRC Institute of Arable Crop Research, Rothamsted Experimental Station, Harpenden, Herts AL5 2JQ, UK

B.A.Whitton
Department of Biological Sciences, University of Durham, Durham DH1 3LE, UK

Abbreviations

ALMV	Alfalfa mosaic virus
Bt	*Bacillus thuringiensis*
CFU	Colony forming unit
CIAT	Centre for Topical Agricultural Research (Colombia)
CMV	Cucumber mosaic virus
CP&E	Crop protection and enhancement
EMBRAPA	Empresa Brasileira de Pesquisa Agropecuaria (Brazil)
EPA	Environmental Protection Agency (USA)
ICARDA	International Centre for Agricultural Research in Dry Areas (Syria)
IITA	International Institute of Tropical Agriculture (Nigeria)
IPM	Integrated pest management
IVR	Inhibitor of viral replication
PGPR	Plant growth-promoting rhizobacteria
PR	Pathogenesis-related
TMV	Tobacco mosaic virus
ToMV	Tomato mosaic virus
VA	Vesicular−arbuscular

CHAPTER 1

An overview of crop inoculation

RODERICK M.MACDONALD

AFRC Institute of Arable Crops Research,
Rothamsted Experimental Station, Harpenden, Herts
AL5 2JQ, UK

Introduction

The purpose of this chapter is to explain how the subsequent ones fit together to form a coherent story. Also, the major relevant subjects which will not be covered in detail in this volume will be discussed briefly.

First, why is crop inoculation of interest? The main reasons for inoculating crops are to attempt to improve crop nutrition and to attempt to control pests and diseases. There is a third, less important reason, the possibility of introducing microbial production of metabolites such as flavourings, drugs and other fine chemicals into plants. In other words, plants can be used as production factories for these compounds. These possibilities are in their infancy at present and are described briefly in the proceedings of the meeting *International Conference on the Present and Potential Industrial Uses of Parasitic and Symbiotic Plant Microorganisms* in *Symbiosis*, 2 (1986).

Crop nutrition

In discussing inoculation which is done to improve crop nutrition, it is convenient to start with microbially mediated nitrogen uptake.

Starting in the 1930s and 1940s, millions of hectares bearing various crops in the USSR were treated with inocula of *Azotobacter* (Rubenchik, 1963). The inoculum was formulated in various ways and was referred to as a bacterial fertilizer, Azotobacterin. The assumption was that, since the bacteria could fix nitrogen, they would do so and benefit the crops. There was no evidence for this. Few of the inoculations were controlled, and a tiny minority of the yields was analysed statistically. This work now seems to have stopped. Another bacterial fertilizer, called phosphobacterin, and consisting of *Bacillus megatherium*, has also been extensively used in Eastern Europe. It allegedly provided crops with phosphate solubilized from the soil pool, but its use has now died out (M.E.Brown, pers. comm.).

Members of the Azotobacteriaceae can be interesting plant inocula under carefully controlled conditions since they are well-known producers of plant hormones (Brown and Burlingham, 1968). This fact is largely responsible for the many reports of argonomically useful stimulation of crop growth by inoculation with these bacteria.

Reproducible stimulation of crops by hormones produced microbially is, however, very difficult to achieve (Beringer et al., Chapter 8). It should be noted that *Azospirillum* and the more recently described *Herbaspirillum* (Baldani et al., 1986) are closely related to *Azotobacter*.

Fixation of nitrogen by rhizosphere bacteria is called 'associative fixation'. This is not a highly evolved, efficient symbiosis. Reduction of N_2 gas to NH_4^+ is energetically expensive (4 g C g^{-1} N fixed in *Azospirillum brasiliense* to a maximum of 174 g C g^{-1} N in *Klebsiella*; Giller and Day, 1985). Only in efficient symbioses is sufficient carbohydrate transported directly to N_2-fixing bacteria to support significant N-fixation. It should be emphasized that non-symbiotic systems make no proven economic contribution in agronomic conditions.

In marginal soils and environments, the combined effects of non-symbiotic N_2-fixing bacteria on plant surfaces and in the soil may be enough to allow a plant to grow where none would grow without biological nitrogen fixation. This may be significant in ecological terms which are relevant in subsistence-level farming, but agronomically it is irrelevant. This controversial subject has been well criticized by Lethbridge and Davison (1983) and Giller and Day (1985).

The *Rhizobium*-legume symbioses are efficient nitrogen-fixing systems in which the value of certain inoculations is beyond doubt. They are considered in detail by Eaglesham (Chapter 3).

The actinomycete genus *Frankia* has been isolated in pure culture from the root nodules of a range of woody perennials (Callaham et al., 1978). The angiosperms involved are not members of the same taxon, and many have little in common. Commercialization of inoculation is rudimentary at present. A very small number of Canadian and American nurseries sell nodulated *Alnus* which is usually used in land reclamation. This group of plants is potentially of great importance, for example *Casuarina* is extensively used in Egypt for sand-dune stabilization. There is also considerable use of *Hyppophae* in stabilizing motorway embankments and coastal sand dunes in this country, and the berries of this plant are used to make a vitamin-C-rich drink in Russia; in the UK several million alders are planted annually (C.T.Wheeler, pers. comm.).

The significance of nodulation is appreciated in many countries, and crushed nodules are often used as inocula. In less well-developed parts of the world there is still ignorance of the value of nodulation. Until a major role is identified in agro-forestry for these symbioses, commercial exploitation will probably progress slowly. The symbioses have been discussed at length by Gordon et al. (1972).

The blue-green algae associate with a wide range of plants and form nitrogen-fixing symbioses. The bacteria involved can often grow and fix nitrogen independently of their hosts. They do not induce the host to form specialized structures, although limited anatomical modification, such as in the coralloid roots of cycads, can occur. All cycads have nodules containing cyanobacteria on their soil-surface roots. It is generally speculated that the ecological significance of these symbioses was in pre-Cretaceous times when cycads were much more common than they now are (Jeffrey, 1987). Commercial exploitation of cyanobacterial symbioses occurs only in the *Azolla*—*Anabaena* partnership which has been successfully used in rice culture for centuries. This subject is discussed by Whitton and Roger (Chapter 7).

The only significant microorganisms to consider in microbially-mediated crop phosphorus uptake are the various mycorrhizal fungi discussed by Stribley (Chapter 4). The fungi may have other roles as well as improving a plant's ability to extract soil phosphorus. The other roles are less well understood and exploited than phosphorus nutrition, and include alterations in a crop's trace-element uptake, drought tolerance and disease resistance.

There is no microbial involvement in uptake of the other major crop nutrient potassium, since this is in plentiful supply in soil and there has been no evolutionary pressure to produce potassium-specific symbioses.

Biological control

The second major reason for crop inoculation is for biological control. Biological control means the limitation or elimination of pests and diseases, in this case by microorganisms. Specific diseases and bacterial and fungal control agents are discussed by Campbell (Chapter 5). Viral crop diseases and strategies derived from plant virology to protect crops are discussed by White and Antoniw (Chapter 6). Viruses can be used as vectors by means of which chosen genes can be incorporated into plants. This is a convenient link with the paper by Beringer, Powell and Poole (Chapter 8) where the genetical aspects of microbial inocula for plants are discussed.

Pest control is obviously closely related to disease control, and some of the same important microbial features apply to both problems. Microbial pest control in crops is the major area which will not be covered in detail in this book. This is to avoid duplication of the published proceedings of a number of recent scientific meetings. The subject was recently reviewed by Wood and Way (1988). There will follow a brief outline of the work which is in progress at present in microbiological crop pest control and it should be remembered that there are also microbial control measures against weeds. Details of these inoculation strategies are available from the specialist publications quoted.

It is impossible to draw the line accurately between crop inoculation for pest control and inoculation of soil for the same purpose. In this sense, we must often consider inoculating the soil−crop−disease complex. Another difference between pest and disease control concerns the mechanism of the microbial effect. Pests are often controlled by specifically pathogenic microorganisms, whereas biological control of disease is more indirect and involves a variety of mechanisms like antibiosis, niche exclusion and plant stimulation.

It is important to consider why we are concerned with microbial pest control. In *Table 1*, chemical and microbial pesticides are compared. It is clear that the reason for using microbial pesticides is that microorgansims have some advantages over chemicals.

Table 2 gives examples of microbial pesticides which are currently used. Others which are under investigation will be briefly mentioned in this chapter. *Table 3* summarizes the properties of a good microbial pesticide; these should be borne in mind in considering specific examples.

Bacterial insect control agents

The bacterium *Bacillus thuringiensis* (referred to as Bt by biotechnologists) was first isolated in 1911 from larvae of the moth *Ephestia kuhniella* in a flour mill in Thuringen,

Table 1. Advantages and disadvantages of microbial versus chemical pesticides[a].

	Chemical	Microbial
Costs/benefits		
R & D	$US20m	$US0.8−1.6m
Market size required for profit	$US40m year^{-1} to recoup development costs, therefore limited to major crops	Markets under $US1.6m may be profitable due to low development costs
Toxicological testing	$US10m	$US0.5m
Patentability	Well established	Still developing
Discovery	Screen 15 000 compounds to identify one product	Rational selection for specific target pests
Efficacy		
Kill	c. 100%	Usually 90−95%
Speed of kill	Rapid	Can be slow
Spectrum of activity	Generally broad	Generally narrow
Resistance	Often develops	Only one known case
Type of action	Can be both preventive and curative	Generally only curative
Safety		
Operator safety	Chemicals can be hazardous	Low operator risk
Environmental impact	Many examples, e.g. accumulation in food chains	Few examples with inundative use of indigenous micro-organisms
Residues	Interval before harvest often required	Crop can usually be harvested immediately after application

[a]Data from F.Paton, pers. comm.

Germany. It is an aerobic spore-former which has a parasporal protein crystal (Beringer *et al.*, Chapter 8) which is very toxic (on ingestion) to certain insects. The specificity of the toxin is high, and lepidopteran larvae are the best-known targets for this control agent. Ingestion of the protein by a susceptible insect leads to cessation of feeding correlated with paralysis of the gut and the mouth parts. Death occurs 30 min to 3 days after ingestion. Deacon (1983) lists a number of target pests against which Bt is a registered product in the USA.

There are at least 30 well-defined varieties of Bt which are differentiated by their H (flagellar) antigens. At present almost all the bacterial insecticides used to control lepidopteran pests are derived from a single strain (HD1) of one serovar.

Bt toxin now accounts for 1.5% of world insecticide sales which in 1986 represented a turnover of $US50−55m. Because of market forces, this figure has varied widely in the last few years (see Lethbridge, Chapter 2).

Table 2. Examples of microbial pesticides[a].

		Target	Crop	Trade name
Insecticides				
Bacillus thuringiensis	B	Various Lepidoptera	Brassicas, trees	Thuricide Dipel
Verticillum lecanii	F	Whitefly	Glasshouse crops	Vertalec Mycotal
Cydia pomonella GV	V	*Cydia pomonella* (codling moth)	Apple, pear, walnut	
Heterorhabditis heliothidis	N	Phorids, sciarids, cecids	Mushrooms, ornamentals	
Fungicides				
Trichoderma viride	F	*Ceratocystis ulmi* (Dutch elm disease)	Elm trees	Binab T
Bactericides				
Agrobacterium radiobacter	B	*Agrobacterium tumefaciens* (crown gall)	Soft-fruit trees, ornamentals	Galltrol
Herbicides				
Collectrotrichum gloeosporioides	F	*Aeschynomene virginica* (Northern joint vetch)	Rice, soyabean	Collego

B, bacterium
F, fungus
V, virus
N, insect parasitic nematode
[a]Data from F.Paton, pers. comm.

In 1977, a new variety (*B.thuringiensis* var. *israelensis*) was found to have activity against mosquitoes (Culicidae) and simulid blackflies (Simuliidae). This was rapidly developed commercially for control of pesticide-resistant blackfly vectors of river blindness in West Africa. More recently, varieties active against beetles have been discovered (F.Paton, pers. comm.). As a microbial pest control agent which does not need to proliferate to be effective, Bt has great advantages over fungal biocontrol agents for which environmental humidity is rarely optimal.

The most exciting developments in Bt technology are taking place in the UK. Burges and Jarret at the AFRC Institute for Horticultural Research, Littlehampton, have produced a patent filed by the Agricultural Genetics Company for a genetically improved strain of Bt. Good control of the commonest moth pest under glass, the tomato moth, can be obtained from this Bt. The less common but resistant cabbage moth survives, however, and this has sometimes prevented the use of Bt. At the Institute for Horticultural Research, Littlehampton, strains of Bt active against the cabbage moth are being selected. These bacteria have sufficient potency against key outdoor pests so that they will have a market profitable enough to attract industrial development.

Table 3. Useful characteristics of a biological control agent.

Characteristic	Persistence of inoculum		(No-inoculum) naturally occurring control agent
	Short-term	Long-term	
1. Rapid colonization of soil	−	+	−
2. Persistence	−	+	+
3. Virulence	+	+	+
4. Predictable control below economic threshold	+	+	+
5. Easy production and application	+	+	−
6. Good storage, without special conditions	+	+	−
7. Low cost	+	+	−
8. Compatibility in integrated control programs	+	+	+
9. Safe	+	+	+

Modified from Kerry (1987).

By the early 1970s, strain selection and improved fermentation conditions had achieved 100-fold increases in potency over the early products. The International Programme for Strain Improvement in Bt (1975) revealed wild strains with 60-fold greater potency than the industrial strain HD1 against certain pests. One of these strains is now being used against wax moths in a product called Certan. It also has good activity against army worms (*Spodoptera* spp.), the larvae of a migratory moth which does great damage to a variety of tropical crops. Importantly, army worms are not controlled economically by HD1.

The strain improvement programme also accumulated a bank of strains of Bt which allowed greater understanding of toxigenicity. The proteins of some strains were found to contain at least three polypeptides whose entomotoxicity differed. The finding that the genes encoding toxin were plasmid-borne, and could be exchanged between some strains of Bt at very high frequency, laid the foundations for genetic manipulation. Using mutagens, bacteria active against a single pest have been obtained. Plasmid-curing studies led to increased understanding of the role of specific proteins in controlling specific pests. This has led to production of recombinants active against particular pests, like the cabbage moth. It is now becoming possible to design bacterial insecticides to be effective against particular pests.

The genes from Bt which code toxigenicity have also now been incorporated into a rhizosphere *Pseudomonas* (Beardsley, 1984). It is hoped that insect pests which attack roots will ingest sufficient modified *Pseudomonas* to effect control. This research has been done by Monsanto in St Louis, but the product is not yet on the market.

The only other significant bacterium in insect control is *Bacillus popilliae*. It is a pathogen rather than being simply toxigenic like *B.thuringiensis*. It is used to control the larval stages of the Japanese beetle in ornamental turf in the USA. Spore dusts are applied to turf, but since the bacterium does not spore *in vitro*, these have to be pro-

Table 4. Comparison of *B.thuringiensis* and *B.popilliae* as microbial control agents[a].

	B.thuringiensis	*B.popilliae*
Pests affected	Lepidoptera (many)	Coleoptera (few)
	Coleoptera (very few)	
Pathogenicity	Low	High
Speed of effect	Immediate	Slow
Role of toxins	Important	Questionable
Current use	Widespread	Restricted (USA)
Method of use	Microbial insecticide	Introduction
Formulation	Spores + toxin	Spores
Production	*In vitro*	*In vivo*
Persistence	Little or none	Marked
Resistance in pest	None reported	Reported but unconfirmed

[a]Data from Deacon (1983)

duced in infected beetle larvae. This makes them expensive and limits applicability of the product. The properties of *B.thuringiensis* and *B.popilliae* in the context of biological control are compared in *Table 4*.

Fungal and viral pest control

Fungal insect control agents are not yet widely commercialized. *Verticillium lecanii* (*Table 2*) was the first fungal insecticide to be successfully commercialized, in the early 1980s, for the control of glasshouse aphids (Aphididae) and whitefly (*Trialeurodes vaporariorum*) in the UK. Mainly for reasons unrelated to the efficiency of the product, it is no longer marketed. Problems of industrial inoculum production and the humidity in the crop canopy are shared with other fungal biocontrol agents.

The fungal order Entomophthorales contains insect pathogens which attack agricultural and horticultural pests as well as insect vectors of mammalian disease. They have been studied for about a century, but are not in commercial use (Wilding and Latteur, 1987). This is because problems of producing fungal inocula and inducing maximal fungal activity on the crop have not been overcome.

The line between crop inoculation for pest control and direct biological control of plant pests is impossible to define clearly. In this latter area, insect viruses have been used, for example to control cotton bollworm (*Heliothis armigera*), pine caterpillars (Lepidoptera) and spruce sawfly (*Gilpinia hercyniae*). Natural virus pandemics are often very important in this area. Recent work on codling moth granulosis virus (*Table 2*) stopped short of a commercial product because of difficulties in producing enough virus, which of course can only be done *in vivo*.

Nematodes

Some more pest control microbiological research which is near to being marketed involves a symbiotic association between the entomopathogenic bacteria *Xenorhabdus* and nematodes in the genera *Heterorhabditis* and *Steinernema*. Experiments are being optimized to control insect pests in the compost in which ornamental plants are grown.

Large quantities of nematodes are produced, using almost microbiological technology. These are introduced to the compost, where they are attracted to susceptible insects. They enter their hosts through the cuticle; the symbiotic bacteria then multiply and are released into the haemocoel. The insect is killed by septicaemia 24–48 h after invasion.

Nematodes can also be important plant pathogens. Since chemical nematicides are very poisonous and kill a wide range of harmless microorganisms, biological control methods for nematodes are of great interest. It has been found for example that some *Streptomyces* spp. produce compounds now called avermectins that are 10–30 times as toxic to nematodes as some classical nematicides (Putter *et al.*, 1981).

Work on predatory nematodes, nematode-parasitic bacteria and nematophagous fungi has not led to large-scale commercial exploitation. Strains of *Arthrobotrys irregularis* and *A. robusta*, which are nematode-trapping fungi, are produced commercially in France (Kerry, 1987) although their usefulness is restricted.

Mycoherbicides

Nor-Am Chemicals Collego (*Table 2*) is the oldest example of a successful mycoherbicide. It is used in the rice and soyabean areas in Arkansas. Several other mycoherbicides are under development and a wide range of chemical and agricultural chemical companies are involved as well as more recent 'biotechnology-based' companies (see Lethbridge, Chapter 2).

Mycoherbicides share some of the advantages and disadvantages of fungal insect control agents, such as specificity and environmental humidity requirements. The possibility exists of using them in combination either with one another or with chemical herbicides. Since their targets are specialized, it is unlikely that they can ever make up more than a tiny fraction of herbicide sales. Their safety, however, is an important plus in the light of the toxicity of some chemical herbicides.

In summary, it is clear that there are at least two major and separate problems confronting researchers on crop inoculation. These are, firstly, establishing that we have a reproducibly useful inoculum, and secondly, establishing that it is commercially viable. The second problem includes formulation and presentation of the product to the farmers. Strictly commercial considerations, as outlined by Lethbridge (Chapter 2), are the only important factors in future commercialization of crop inoculation. Other relevant criteria are important only in so far as they contribute to commercial decisions on the future use of products derived from microorganisms.

References

Baldani,J.I., Baldani,V.L.D., Seldin,L. and Dobereiner,J. (1986) Characteristics of *Herbaspirillum seropedicae* gen. nov., sp. nov., a root-associated nitrogen fixing bacterium. *International Journal of Systematic Bacteriology,* **36**, 86–93.
Beardsley,T. (1984) Genetic pesticides, *Nature,* **312**, 686.
Brown,M.E. and Burlingham,S.K. (1968) Production of plant growth-substances by *Azotobacter chroococcum*. *Journal of General Microbiology,* **53**, 135–144.
Callaham,D., Del Tredici,P. and Torrey,J.G. (1978) Isolation and cultivation *in vitro* of the Actinomycete causing root nodulation in *Comptonia. Science,* **199**, 899–902.

Deacon,J.W. (1983) *Microbial Control of Plant Pests and Diseases (Aspects of Microbiology 7)*. Van Nostrand Reinhold, Wokingham.

Giller,K.E. and Day,J.M. (1985) Nitrogen fixation in the rhizosphere: significance in natural and agricultural systems. *Ecological Interactions in Soil. British Ecological Society, Special Publication,* **4**, 127–147.

Gordon,J.C., Wheeler,C.T. and Perry,D.A. (1972) *Symbiotic Nitrogen Fixation in the Management of Temperate Forests*. Forest Research Laboratory, Oregon State University, Corvallis.

Jeffrey,D.W. (1987) *Soil–Plant Relationships. An Ecological Approach*. Croom Helm, London and Sydney.

Kerry,B.R. (1987) Biological control. In R.H.Brown and B.R.Kerry (eds), *Principles and Practice of Nematode Control in Crops*, Academic Press, Sydney.

Lethbridge,G. and Davison,M.S. (1983) Root-associated nitrogen-fixing bacteria and their role in the nitrogen nutrition of wheat estimated by ^{15}N isotope dilution. *Soil Biology and Biochemistry,* **15**, 365–374.

Putter,I., MacConnel,J.G., Preiser,F.A., Haidri,A.A., Ristich,S.S. and Dybas,R.A. (1981) Avermectins: novel insecticides, acaricides and nematicides from a soil micro-organism. *Experientia,* **37**, 963–964.

Rubenchik,L.I. (1963) *Azotobacter and its use in Agriculture*. Academy of Sciences of the Ukrainian SSR. Microbiological institute im. D.K. Zabolotnyi. Israel Programme for Scientific Translations, Jerusalem. (From the Russian text published in 1960.)

Wilding,N. and Latteur,G. (1987) The Entomopthorales—problems relative to their mass production and their utilisation. *Med. Fac. Landbouw Rijks Univ. Gent.,* **52**, 159–164.

Wood,R.K.S. and Way,M.J. (1988) Biological control of pests, pathogens and weeds: development and prospects. *Philosophical Transactions of the Royal Society of London, Series B,* **318**, 109–376.

CHAPTER 2

An industrial view of microbial inoculants for crop plants

G. LETHBRIDGE

Physiology and Biochemistry Division, Shell Research Ltd, Sittingbourne Research Centre, Sittingbourne, Kent ME9 8AG, UK

Introduction

The concept of microbial inoculation of crop plants to increase yield goes back about 100 years and people have been trying to make money out of it for the last 50 years. Despite this long history, the global markets for microbial pesticides and so-called microbial fertilizers have been, and still are, insignificant when compared to the corresponding markets for the equivalent chemical products. The reason for this is that historically the performance of microbial products has fallen well below that achievable with chemicals under most circumstances. Consequently, although most of the major participants in the agrochemical business have shown an interest in microbial inocula for crop protection and enhancement (CP & E) at one time or another, commercial development of the few products available has been left primarily to small companies specializing in microbial inoculants, or to larger companies whose major interest is not agrochemicals.

However, in recent years the climate inside the agrochemical industry has been changing, and microbial CP & E is back on the agenda. This renewed interest is due, in part, to an optimism that recent technological advances offer solutions to past problems with microbial CP & E agents, while being spurred on by an uncertainty concerning a number of issues on the horizon with regard to chemical products.

In this contribution, some of the challenges that the agrochemical industry is currently facing and is likely to face in the future will be dealt with. The following points will be considered: the potential attraction of microbial CP & E agents to the industry; the current commercial status of microbial CP & E products; the barriers to commercialization of new ones; the options open to the industry with regard to microbial CP & E; and finally some generalized market niches for these products will be identified.

In the chemical context, CP & E is taken to include pesticides and plant growth regulators, but not fertilizers. The discussion will concentrate on microbial crop protection, with only passing reference to microbial fertilizers.

Challenges in the agrochemical industry

In 1986 the size of the global market for chemical CP & E agents was around $US17.4b

Table 1. The size of the global market for crop protection and enhancement chemicals in 1986 by product line.

Product line	Market value ($US b)	Market share (%)
Herbicides	7.60	44
Insecticides	5.45	31
Fungicides	3.25	19
Plant growth regulators	1.10	6
Total	17.40	100

Data from Wood-Mackenzie (1987).

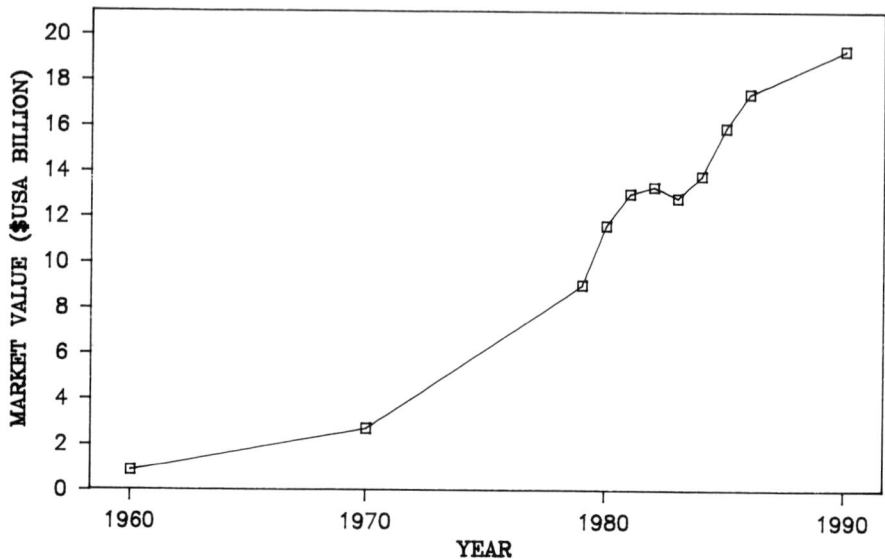

Figure 1. Development of the global CP & E market (1960–90).

(Wood-Mackenzie, 1987); the product line split is shown in *Table 1*. Historically, the CP & E market has grown on average by around 10% per annum since the 1940s, but more recently growth has started to slow down (*Figure 1*). Wood-Mackenzie (1987) estimated that real growth over the period 1988–90 will average 3% per annum, thus valuing the 1990 global market at $US19.3b. There are two reasons for this slowing down in the growth rate of agrochemical demand. First, population growth in the developed world is static, and secondly, farm incomes are declining as over-production leads to a reduction in farm-gate prices for many commodity products. In certain parts of the world, as incomes have fallen, there has been a reduction in the area of land planted, and hence a reduction in the demand for CP & E products. In the USA, the area planted to food and feed grains in 1987 was estimated to have fallen by 10% compared to the previous year (Wood-Mackenzie, 1987). The share of the global market for all CP & E agents combined attributed to the USA has fallen from 34% in 1984 to 26% in 1986, and is likely to continue falling at least in the short term (Wood-Mackenzie, 1987). The reduction in certain market sizes has led to fierce competition

An industrial view of inoculants

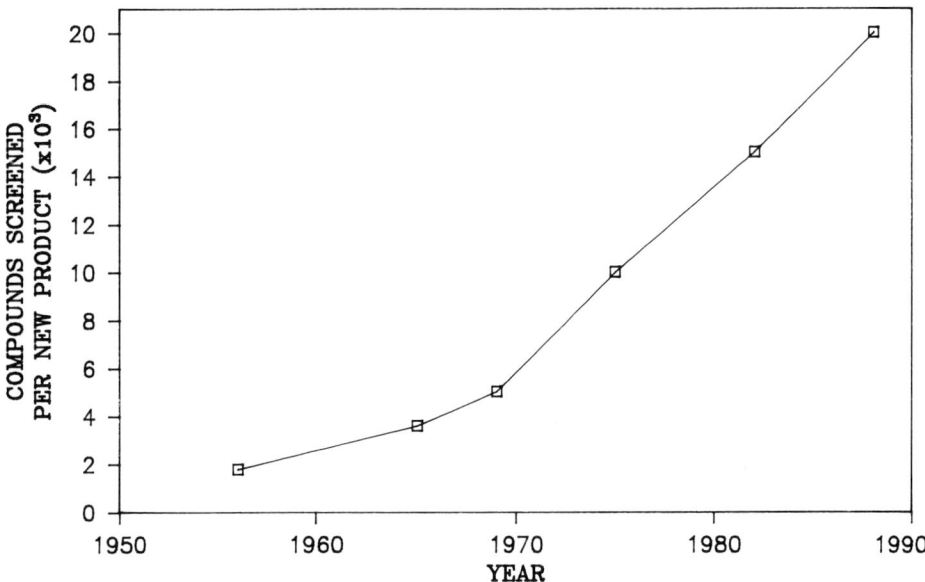

Figure 2. Statistical chance of discovering a commercial CP&E agent over the last 30 years.

in pricing, something which has been evident for commodity products for some time but which is now starting to have its effect on proprietary compounds, forcing some of them on to the downward spiral towards commodity status. This problem is compounded for some companies as their products' patents expire, although patent restoration would obviously help alleviate the problem. Companies which have recently faced or are about to face this situation include Eli Lilly (Treflan) and Monsanto (Lasso and Roundup), all the products in question being herbicides.

Long-term survival in the agrochemical business is dependent on the discovery of new compounds, which requires a significant commitment to research and development. However, the law of diminishing returns means it is becoming harder to find new commercial products (*Figure 2*). The current discovery rate is estimated to be in the region of 1 in every 20 000 compounds screened (and rising), compared to 1 in every 5000 compounds screened 20 years ago (Rowe and Margaritis, 1987). Coupled with this, the rate of introduction of new molecules has fallen from an average of 9 per annum in the 1970s to an average of 7 per annum in the 1980s. In addition, it is becoming increasingly difficult to recover the considerable costs of research and development through the introductions of new products. Wood-Mackenzie (1987) has calculated that the cost of bringing a new molecule to the market place can be as high as $US80m. They arrived at this average for the industry by making the following assumptions: the time between initial synthesis and commercial launch is 10 years; the average R&D spend for the industry is 8% of sales; new product R&D accounts for 60% of the total R&D budget; the industry launches an average of 7 commercially significant products per year.

The contribution of registration costs to this figure can be as high as $US20m (Agrow, 1985), with the possibility that they will increase. A recent survey of groundwater analyses in the USA revealed the presence of most commercially available fungicides (Carsel and Smith, 1987). This has led to calls for stricter registration testing for new chemicals and re-registration testing for some existing products (Agrow, 1985).

Development of resistance to insecticides and fungicides may shorten the commercial life of these products, particularly systemic ones such as the benzimidazole fungicides.

Against the background described above, how does the agrochemical industry view the opportunities for microbial CP & E agents?

Potential attractions of microbial CP & E agents to the agrochemical industry

Reduced discovery costs

The most fundamental difference between screening chemicals and microorganisms for CP & E activity is that microbial screening permits a more selective approach. Microorganisms need not be screened randomly through a multiplicity of screens designed to pick up as wide a range of biological activity as possible. Instead, the desired target activity is selected in advance of screening, and with this in mind, specific environments are identified where microorganisms with the desired activity are likely to be found. Thus, a potential mycoherbicide could be isolated from diseased weeds identified in the field (Templeton and TeBeest, 1979). Antagonists of soil-borne root pathogens can be isolated from disease-suppressive soils (Cook and Baker, 1983). Experience shows that it is relatively easy to isolate from such environments microorganisms which show good biocontrol potential in laboratory or glasshouse screens, without having to play an enormous numbers game. Superficially, therefore, it appears that the discovery costs of a microbial CP & E agent should be smaller than those for a chemical. This is certainly the case for products already on the market such as insecticides based on wild-type strains of *Bacillus thuringiensis* and the mycoherbicide *Colletiotrichum gloeosporioides* (Collego), which have cost an estimated $US1m and $US2m, respectively, to develop prior to registration (Lysansky, 1984; Templeton, 1986). The extent to which this cost advantage can be maintained with new products will depend on the amount of effort required for strain improvement either by traditional or recombinant DNA techniques, or for the translation of glasshouse potential into performance under end-user conditions. The latter is a major problem where microbial CP & E agents are being developed for use under field conditions, as opposed to protected environments. This is not a trivial problem and will be re-examined later.

Reduced registration costs

In the United States all new CP & E products must be registered with the Environmental Protection Agency (EPA) to prove their safety before going on sale. This registration is dependent upon the new product satisfactorily passing a thorough toxicological and environmental testing programme. The EPA refer to microbial CP & E agents as 'biorational pesticides', a term which also includes certain (but by no means all) natural products used in this context, such as insect pheromones. They believe that CP & E agents based on naturally-occurring microorganisms indigenous to the area of intended use generally pose fewer environmental concerns than synthetic chemical agents. Consequently, a new approach to safety testing of biorational pesticides has been

Table 2. Summary of data required by US Environmental Protection Agency (EPA) for the registration of a crop protection agent based on a microorganism indigenous to the United States.

Environmental	Mammalian toxicology
1. Phytotoxicity on the target crop	1. Oral
2. Phytotoxicity on specified non-target crops	2. Eye
3. Non-target insects	3. Dermal
4. Freshwater, estuarine and marine habitats	4. Inhalation
5. Wild mammal testing	5. Injection
6. Avian testing	6. Hypersensitivity
	7. Tissue culture

Additional information
1. Strain identification
2. Mode of action
3. Ecology of microorganism after introduction

Data from EPA (1980).

proposed (EPA, 1980), based on a 3-tier system, where the maximum challenge is encountered in tier-1. EPA argue that any undesirable effects of microbial products should be detected in tier-1 of their protocol (*Table 2*). Any microbial product satisfactorily passing tier-1 testing is exempt from further toxicological and environmental testing.

By their approach, EPA are effectively trying to encourage more companies to develop microbial CP & E products, because this initiative dramatically reduces the registration costs of a new microbial product compared to those of a synthetic chemical, which can be as high as $US20m (Agrow, 1985). In contrast, Markle (1983) estimated that tier 1 testing for the registration of a microbial product could be achieved for $US141 000 in 1983. Taken in isolation, the current cost of registering a wild-type microbial CP & E agent must be considered a major incentive for a company to move in this direction.

Reduced lead time

The reduced testing requirements for microbial CP & E agents not only reduce registration costs, they also significantly reduce the lead time between initial discovery and product launch. On average this takes about 6−7 years for a synthetic chemical, with registration testing accounting for two-thirds of this time. In contrast, it has been claimed that a microbial product could be on the market 3−5 years after the target has been identified (Templeton, 1986). This estimate may be over-optimistic, but the important point is that the rate-limiting factor is the rate of progress in research and development, rather than registration testing.

Smaller markets viable

The costs associated with bringing a new chemical CP & E agent to the market place are so large that these products must be targeted towards markets large enough to permit

Table 3. Leading sectors in the crop protection market in 1986.

Crop	Crop protection agents	Market value ($US m)
Maize	Herbicides	1585
Fruit and vegetables	Fungicides	1475
Soyabean	Herbicides	1450
Cotton	Insecticides	1405
Fruit and vegetables	Insecticides	1385
Wheat	Herbicides	1080
Rice	Insecticides	900
Rice	Herbicides	765
Fruit and vegetables	Herbicides	700
Rice	Fungicides	550
Maize	Insecticides	485
Wheat	Fungicides	465
Sugar beet	Herbicides	415
Cotton	Herbicides	360

Data from Wood-Mackenzie (1987).

a satisfactory return on investment within a reasonable period of time. This restricts the choice of targets to those affecting production of the world's major crops (*Table 3*). A new opportunity should have projected annual sales of at least $US10–20m to be commercially viable, but this will of course be influenced by the size of the margin. In contrast, the considerably lower costs associated with putting a microbial product on the market means that theoretically much smaller markets become viable. I use the word theoretically, because in a large chemical company, management may not differentiate between chemical and microbial products, so that the overheads for new product R&D may be shared equally between both types of product. Under these circumstances microbial products will appear to be more expensive to develop than if they had been considered independently of chemicals, and the smaller market opportunities will be lost. One way round this problem would be for the large company to set up a separate small company and hence create the environment where small projects are not killed by the massive overheads of chemical development. However, there will still be limits to market size and viability under this scenario. Templeton (1986) has estimated the cost of putting the mycoherbicide Collego (for the control of northern joint-vetch in rice) on the market as $US2m (in money of the day). Approximately 6% of the rice crop in Arkansas was sold at a discounted rate between 1980 and 1982 owing to the presence of the black seeds of the weed (Smith, 1986). The cost of this loss has been estimated at $US4m per annum (Charudattan, 1985). Assuming a favourable cost:benefit ratio of 1:4, this values the market for control of northern joint-vetch at $US1m per annum. In 1983 some 6000 hectares were treated at a cost to the farmer of $US22 per hectare, thereby valuing the Collego market at around $US132 000 per annum. Given these costings, this technically successful project will never break even, let alone generate a satisfactory return on investment. Despite these economic problems, the Collego project has been very useful as a learning exercise.

Integrated pest management

The middle ground between chemical and microbial CP & E is integrated pest management (IPM), where a combination of both approaches is used to solve a particular problem. Indeed this must be the way forward, if microbial products are ever to have a significant impact on crop production. In the short term, this approach could benefit existing chemical products. The aim would be to find a microbial CP & E agent whose performance may not be good enough to form the basis of a commercial product on its own but, when combined with an existing chemical which has its own problems, forms an effective, commercially viable product. It ought to be relatively easy to select for strains of the microorganisms under test which are resistant to the chemical with which they are to be applied. The chemical may benefit from this strategy by a reduction in the amount required to achieve effective control. This may widen the number of applications for which the chemical can be registered and thereby increase its revenue-earning potential. In addition, this mixed strategy should help reduce the risk of resistance.

Resistance

One of the often-quoted advantages of microbial CP agents over chemical products is that the former are less prone to problems of resistance than the latter. This may well be the case when some of the modern site-specific chemicals are compared to microorganisms which are antagonistic through a combination of mechanisms. However, there is no reason to suppose that microorganisms which are antagonistic via the activity of a single product (for example an antibiotic or toxin) will be any less prone to resistance problems than synthetic chemicals. It is a fallacy to claim that resistance to *B.thuringiensis* will not occur because it is a naturally occurring bacterium (Crull, 1985). Indeed, resistance to Bt toxin began to appear in the USA in the very year that this statement was made (McGaughey, 1985).

Versatile manufacturing plant

Although fermentation is a more expensive means of manufacturing than chemical synthesis (see p.20−21), a fermentation plant does have the advantage of being more versatile than a chemical plant. Thus, whereas at present a chemical plant tends to be restricted to the manufacture of a specific chemical, a fermentation plant can be used to produce a variety of products.

Environmental acceptability

The 'natural' tag associated with CP & E products based on non-engineered, indigenous microorganisms is likely to make them environmentally more acceptable than chemicals in the eyes of 'green' pressure groups and the general public. The industry too is concerned that its products are environmentally acceptable and goes to extreme lengths to test its products before they reach the marketplace. Consequently, the 'natural' tag alone will not be sufficient to move the industry in the direction of microbial CP & E agents. Before this happens, these products must also demonstrate performance and commercial viability.

Having discussed the incentives for industry to get involved with microbial CP & E agents, the current commercial status of these products will be briefly considered.

The current commercial status of microbial CP & E products

The size of the global market for microbial CP & E products (excluding *Rhizobium*) is approximately $US25m (0.14% of the size of the chemical market) with products based on *B.thuringiensis* accounting for almost 90% of the market (*Table 4*). This value is less than the often-quoted figure of $US30−40m, owing to a drop in the market price of almost 50% for insecticides based on *B.thuringiensis* over the period 1984−85 (Biotechnology Affiliates, 1987). Two-thirds of the global market for *B.thuringiensis* is in North America (*Table 5*) where it is used primarily in forestry (*Table 6*). Its specificity and rapid inactivation have severely restricted its use on other crops, but its use in forestry is favoured through environmental pressures (Burges, 1986). In addition, a small amount of defoliation can be tolerated in the forestry situation, so

Table 4. The global market for microbial inocula.

Product	Market value ($US m)
Bacillus thuringiensis insecticides	22
Other microbial pesticides	3
Bradyrhizobium japonicum	18
Other microbial fertilizers	< 5

Data from Biotechnology Affiliates (1987); Jutsum (1987); and personal communications.

Table 5. The global market for *Bacillus thuringiensis*.

Geographic region	Market value ($US m)
North America	14.0
Far East	2.5
China	2.0
Central and South America	1.0
Middle East and North Africa	1.0
Rest of Africa	1.0
Australasia	0.2
Western Europe	0.1
Total	21.8

Data from Biotechnology Affiliates (1987).

Table 6. The use of *Bacillus thuringiensis* in North America.

Market sector	Market value ($US m)
Forestry	
Canada	4.4
USA	3.7
Vegetables (lettuce and brassicas)	1.3
Soyabeans, tobacco, alfalfa	0.4
Grapes	0.2
Blackfly control	2.5
Mosquito control	1.5
Total	14.0

Data from Biotechnology Affiliates (1987).

the lower efficacy of *B.thuringiensis* compared to chemicals is acceptable here. Why, despite almost 70 years' work in this field, is the contribution of microbial products to the CP & E industry so small? It has been claimed by some that if the same amount of money that was spent on research into chemical CP & E agents over this period had been spent on research into biological CP & E agents, then the contributions would be reversed (Van den Bosch, 1980). This is extremely unlikely, although one would be surprised if, under this hypothetical scenario, microbial products did not account for a greater proportion of today's market than they do currently. Most large companies with an agrochemical business, and some without, have at one time or another become involved with microbial CP & E agents. Several have put products on the market, although with varying degrees of commercial success *(Table 7)*. Those based on *B.thuringiensis* have proved most successful. It is interesting to note that only two of these companies, Sandoz and Schering (through their subsidiary Nor-Am acquiring the Collego business from Upjohn), are major participants in the agrochemicals business. The most successful company in this area (Abbott) is better known for its pharmaceuticals business.

Barriers to the commercialization of microbial CP & E agents

In order to understand why the contribution of microbial inoculants to today's agrochemical business is so small, it is necessary to analyse the requirements of a new commercial CP & E product. This will help to put the industry's concerns about microbial inoculants as commercially viable CP & E agents into perspective. An increased awareness of these problems should aid in the development of the right types of microbial products and increase their chances of commercial success.

Table 7. Commercialization of microbial inocula by multinational companies.

Company	Microorganism	Product type	Status
Abbott	*Bacillus thuringiensis*	Insecticide (Dipel, Vectobac)	Continuing
	Hirsutella thompsonii	Mycoinsecticide (Mycar)	Discontinued
	Phytophthora palmivora	Mycoherbicide (Devine)	Continuing
Sandoz	*B.thuringiensis*	Insecticide (Thuricide, Certan Teknar)	Continuing
	Heliothis nuclear Polyhedrosis virus	Insecticide (Elcar)	Discontinued
Tate & Lyle	*B.thuringiensis*	Insecticide (Skeetal, Biobit)	Sold (Novo)
	Verticillium lecanii	Mycoinsecticide (Vertalec, Mycotal)	Sold (Novo)
Upjohn	*Colletotrichum gloeosporoides*	Mycoherbicide (Collego)	Sold (Ecogen)

To be a serious candidate for commercialization, a new CP & E agent must satisfy the following requirements.

(i) There must be customer demand for that type of product. More often than not this will initiate the research (business pull), but there is also scope for a degree of research push, when customer demand must be created.
(ii) As discussed earlier, the market size has to be large enough to permit a satisfactory return on investment within a reasonable period of time.
(iii) Broad-spectrum activity is a prerequisite for maximizing market size.
(iv) Manufacture must be achievable in a cost-effective manner.
(v) Performance must be both high and reliable.
(vi) The persistence of the compound in the environment should be long enough to achieve the desired effect, without the need for regular application which may be uneconomic; but conversely it must eventually be biodegradable so as to pose no threat to the environment and to ensure repeat sales.
(vii) The product should be free from toxicological problems.
(viii) The end-user formulation must have a minimum shelf-life of 2 years at room temperature, be easy to handle and insensitive to abuse. It must be stable over the range $-5°C$ to $+30°C$.
(ix) If the new product is aimed at a market where products already exist, it will have to be differentiated from the competition on the grounds of performance, cost or ease of handling.

In the light of the above list, what are the industry's concerns about turning microbial CP & E agents into commercial products?

Commercial viability

The question of market size has already been considered, but there are other factors which affect commercial viability, such as selling price and production costs. The selling prices of some commercial microbial CP & E products are given in *Table 8*, along with a range of other fermentation products for comparison. The end-user price of CP & E products (in terms of costs per hectare per season) is calculated from the value of the increased yield to the farmer divided by 5 (assuming a cost−benefit ratio of 1:5). If more than one treatment is required per season to achieve the yield increase, then the cost per treatment is derived from the cost per season by dividing by the number of treatments. Alternatively, if there are existing products on the market, which is usually the case, then the price of these may determine the target selling price for the new product. Insecticides based on *B. thuringiensis* are commodity products with narrow profit margins. Severe market competition over the period 1984−85 resulted in a 50% drop in the market price and the eventual demise of 2 out of the 4 main producers. The selling price quoted for *Bradyrhizobium japonicum* is based on the Agricultural Genetics Company's product sold in Europe.

Details on production costs are much harder to obtain: no one wants to divulge production costs to competitor companies or customers. Production costs will be very sensitive to reductions in viable count at any stage of the production process. Consequently, microbial inocula will be a new challenge for the biochemical engineer, since fermentation and downstream processing must be optimized on maintenance of

Table 8. The selling price of some microbial CP & E products compared to some other fermentation products.

Fermentation product	Selling price US$ per hectare	$US per tonne
Ethanol		500
Yeast biomass		1 000
Citric acid		1 600
Lactic acid		2 200
Monosodium glutamate		2 500
Lysine		4 000
Xanthan		10 000
Bacillus thuringiensis	8	15 000
Penicillin		30 000
Riboflavin		50 000
Bradyrhizobium japonicum	18	72 000
Bulk industrial enzymes		100 000
Collego (mycoherbicide)	22	220 000
Cephalosporins		6 000 000
Vitamin B_{12}		8 000 000

Treatment costs (per hectare) are derived from Rowe and Margartis, 1987 (*Bacillus thuringiensis*); Elson's Seeds Ltd, Spalding, Lincs, UK (*Bradyrhizobium japonicum*); and Templeton, 1986 (Collego). Data for other fermentation products from Hacking (1986).

viability, not specific product stability as is the case with products currently produced by fermentation. An order of magnitude drop in viability at any stage will result in a tenfold increase in production costs.

Reliability

In the past, microbial CP & E agents have been notoriously unreliable. This is certainly the case for many examples in the open literature and from the industry's experience with them. This variability is not just restricted to field experiments, but is also relatively common when the same strain is tested in different laboratories (see Campbell, Chapter 5). Clearly, microbial CP & E agents are very sensitive to environmental change and perhaps it is too much to expect otherwise, at least in the field, where it is not uncommon for a microbial inoculant to fail in 3 out of every 4 trials. However, it is worth pointing out at this stage that chemical CP & E products are not immune to failure due to vagaries of the environment and it is not unusual to have field trial failures with chemicals which will eventually be commercialized. If a microbial inoculant is being targeted towards a market where competing chemicals already exist, it is useful to include some chemical treatments in the inoculant trials for comparative purposes. In addition to environmental regulation of expression of activity, variability of performance may also result from loss of activity during manufacture or after introduction to the environment, although problems of this nature should not be insurmountable.

Efficacy

Although antagonistic microorganisms can reduce pest incidence by more than 95%, this is the exception rather than the rule, and in general they tend to be less effective than chemicals. It is not unusual for microbial agents to achieve less than 50% control. CP & E products are sold on the basis of improvements in crop quality and quantity, that is, a reduction in pest incidence and increased yield. It may be possible to reduce pest incidence by 50%, but to have no effect on the final yield of the crop. From this discussion it will be apparent that measurements of reduction in pest incidence in the absence of any yield data are of limited value from the point of view of an economic evaluation of a project.

A question frequently asked concerns the extent to which crop yield must be increased before a microorganism becomes a serious contender for commercialization. However, it is not possible to generalize in this manner, because increased yield is only one of the factors taken into account in the economic assessment; other factors include the value of the crop and the manufacturing costs of the product.

Another aspect of the question of efficacy is concerned with potency; that is, can they deliver the required effect at a cost-effective dose? This is related to application rate, which will be considered later.

Activity spectrum

In general, microbial CP & E agents have narrow activity spectra. Species specificity is very attractive from the environmental point of view, but in most situations the crop will require protection from more than one weed, fungus or insect species, and it would be uneconomic to treat each with a separate product.

Mode of action

With a few exceptions (such as *B.thuringiensis*), microbial CP & E products are slow-acting. The mycoherbicide Collego takes 4–5 weeks to kill northern joint-vetch (Templeton and Greaves, 1984). By comparison, weeds treated with the slow-acting chemical herbicide glyphosate show serious symptoms in about 10 days (Hassall, 1982). However, slow action may be no bad thing, so long as there is no yield penalty, although it may result in market resistance from farmers used to the rapid action of chemicals. The major problem lies in trying to control pests which penetrate the plant rapidly. This is asking a tremendous amount from a microbial pest control agent, and clearly chemicals have the upper hand here. One important consequence of slow action is that microbial insecticides and fungicides can only be really effective as preventive, not curative, treatments. So they must either be used as an insurance policy, or in conjunction with accurate forecasting.

Duration of control

The duration of pest control may be too short or too long. As a consequence, the product may be uneconomic because too many applications per season have to be made under the former circumstances, or because sales are too infrequent under the latter. On cotton, the half-life of products based on *B.thuringiensis* is no more than 2 days; consequently, their use on this crop is not economic (Andrews *et al.*, 1987). Abbott's mycoherbicide Devine (*Phytophthora palmivora*) for the control of milkweed vine in Florida citrus orchards suffers from the second problem, weed control persisting for several seasons after one application of the soil-borne pathogen (Kenney, 1986).

Formulation

Formulation converts chemicals, or in our case microorganisms, into products. Its importance should not be overlooked, since it can have a marked effect on product performance. Chemical formulations have set high standards with regard to long shelf-life, ease of use and resistance to abuse, and microbial formulations will be expected to match them. Microbial products will have to behave like their chemical counterparts. Loss of viability during formation must be kept to a minimum. A drop in viability of one order of magnitude will increase production costs by 10 times. A shelf-life of 2 years at room temperature and stability over the range $-5°$ to $+30°C$ may seem demanding requirements for some types of microorganism, and this is probably a good enough case for not pursuing these in the first place. Products which have a short shelf-life are unacceptable to the producer on the grounds of economics and to the customer on the grounds of inconvenience. One of the problems with Devine is its very short shelf-life, so that it has to be made to order each season (Kenney, 1986).

Application

One advantage of formulating microorganisms like chemicals is that they can then be applied using standard machinery. This is crucial: no one is likely to put on the market, let alone buy, a specialized piece of equipment for applying a new type of unproven product. Ease of handling during preparation for application is also important. This may be one of the reasons why the acceptance of Collego was poorer than expected, since the product consists of three packages: (i) activated charcoal for cleaning up chemical residues from application equipment; (ii) dried spores mixed with an inert bulking agent; and (iii) a rehydrating agent vital to ensure good spore viability (Bowers, 1986). This is not the most straightforward of preparation regimes.

There should be as few restrictions as possible on the timing of application. Microbial CP & E agents are only likely to work under conditions of high humidity. This can severely limit the opportunities for a foliar active product such as a mycoherbicide or microbial insecticide. Restricting the timing of application to dawn and dusk or overcast days, besides being inconvenient, may not be sufficient for a slow-acting pathogen, and is a good reason for choosing rice as a target crop for these types of products. Indeed, it is recommended that rice be flooded and soyabean irrigated immediately prior to the application of Collego to ensure the best results (Templeton and Greaves, 1984). Humidity is less of a problem when microbial products are to be used in the soil (except perhaps in very dry climates). This is a strong case for concentrating on soil and seed applications for microbial CP & E, although delivering the agent to the required site can be more of a problem in this case than when foliage is the target.

Application rates must match those being used for chemicals and these are falling all the time. Farmers are now used to applying from the low hundreds down to even tens of grams of active ingredient per hectare in 50–400 litres of water (or 1–5 kg of bulking agent for granular application) of the new generation of chemical CP & E products. Application rates quoted for glasshouse experiments or microplot field experiments with microbial CP & E agents in the open literature frequently equate to the equivalent of tonnes per hectare (Lewis and Papavisas, 1987)! Admittedly 99.9% of this material is often inert formulating material (peat, bran or alginate), but application at such high rates is uneconomic. Clearly in such cases the formulating strategy needs to be rethought.

Patents

Industry requires some form of protection for its discoveries as a means of ensuring a reasonable return on its investment in research and development. The agrochemical industry is used to seeking patent protection on the discovery of new active molecules. The problem with patenting is that in return for sole rights to exploit an invention for a fixed period, the invention must be divulged in the patent specification. An alternative to patent protection is to keep an invention as a trade secret. This approach is often used in the food industry where exclusivity beyond the lifetime of a patent is required and the chances of a competitor discovering a secret are considered slight. It is unlikely that the Coca-Cola corporation would be as successful as it is today if it had patented its famous product. Although the situation with regard to patents on genetically engineered microorganisms looks promising (Crespi, 1986), wild-type organisms are not patentable. Clearly, some form of protection is required for wild-type products to prevent a competitor buying some of the product and isolating the strain from it. He is then in a position to start manufacturing the product and attacking the market built up by the original producer. A large company could easily have problems with small companies under these circumstances, since the lower overheads of the latter may enable them to sell the product at a discounted price. Trade secrecy is unlikely to be sufficient to prevent these events, so what are the alternatives? Patenting the formulation is a possibility, but not a particularly attractive one, since it will not stop competitors using an alternative method to formulate the organism. If the formulation is good, a producer may not want to patent it, but may decide on trade secrecy for the reasons discussed above. The best option may be to try to obtain a patent on the use of the microorganism for the particular application in question.

Registration

The concern over registration is not with wild-type organisms which are seen as a positive advantage of microbial CP & E agents over chemicals. The use of genetically engineered strains, however, (see Beringer *et al.*, Chapter 8) must balance the expected benefits to society resulting from their use against the potential risks of introducing recombinant organisms into the environment. It is not in industry's interests to develop a product whose risks are perceived to outweigh its benefits. At present, regulatory authorities on both sides of the Atlantic are feeling their way with regard to risk assessment. A West German Commission of enquiry into the subject has recommended a ban on the release of genetically engineered viruses and a five-year moratorium on the release

An industrial view of inoculants

of all other recombinant microorganisms (Ager, 1987). In Demark, an Act of Parliament has to be passed before a non-indigenous microorganism can be released into the environment. The concern of industry at the moment is that a regulatory framework which is both scientifically sound and economically sensible is produced, and that there is a risk assessment strategy in which society can have confidence (Poole, 1987). Without these tools, industry will not develop microbial CP & E products based on genetically engineered strains, and the benefits will be lost.

Image

The history of microbial inoculant use in agriculture is littered with products of doubtful efficacy, particularly in the field of so-called microbial fertilizers or soil conditioners. Some were produced on a large scale, such as Azotobacterin and Phosphobacterin in the Soviet Union, but the majority were and in some cases still are produced by small companies specializing in microbial inoculants and enzymes for a wide range of applications. These products typically contain a pot-pourri of microbial species (possibly as many as 20) capable of fixing nitrogen, solubilizing phosphate, degrading organic matter constituents found in the soil, and producing plant growth regulators in laboratory culture. However, when rigorously tested, none has been satisfactorily shown to have any significant effect on crop growth in soil, because these types of organisms are ubiquitous in soils and the functions they carry out are limited by factors other than population size in the soil.

The claims for some of these products are quite fantastic, and it is easy to see how the customer, particularly one who has a leaning towards reduced chemical inputs, might be taken in. The attraction of products of this type to the producer is that, because they are labelled as fertilizers, they are not subject to any registration requirements. In addition, it is very hard, if not impossible, for the farmer to know whether the product has worked. As a result there is a danger that all microbial inoculants will be tarred with the same brush, so an element of user and adviser education will be required to convert the sceptical. The saving grace for microbial crop protection is that, for many of the products (but not necessarily all of them), it will be obvious to the eye when they have worked.

Industrial strategies for microbial CP & E

Having summarized the case for and against microbial CP & E agents as perceived by the industry, we may ask what options are open to an agrochemical company faced with deciding whether or not to become involved in this area. The easiest option is to do nothing. The reasons for this may be a belief that the subject is being oversold, that there is nothing in it, that the bubble will burst soon enough and there will be no threat to the existing chemicals business. Secondly, the opportunities for microbial CP & E may not be perceived. Finally, the only market opportunities that can be identified may be unattractive, owing to a conflict of interest with existing products, or their small size making them uneconomic under current company policy.

If the decision is to become involved, this can be done at different levels of commitment, ranging from simply keeping a watching brief on developments in the subject, to assessing third-party products from small biotechnology companies looking for a good marketing network, through to setting up in-house research programs. The reasons for doing this might be that there is a conviction that the subject is advancing and that commercially viable products will be forthcoming, coupled with a desire to be at the forefront. There may be uncertainty about the future of the subject, but a concern about not getting left behind if it really takes off. There can also be political pressure through the banning of chemicals: for example the removal of 2,4,5-T from the market should help the sales of Collego (Templeton, 1986). Finally, the company may have spare fermentation capacity which could be used for microbial CP & E products.

Niches for microbial CP & E products

Once a producer has taken the decision to get involved in microbial CP & E, where might the market opportunities be? Given the limitations of these products compared to chemicals, it makes sense to avoid the markets where there are already effective chemicals, since, all else being equal, with current technology, microbials cannot compete with chemicals in terms of performance. However, there are still niches where opportunities for microbials exist, for the following reasons.

(i) There are no effective chemicals.
(ii) Resistance to chemicals is becoming a serious problem.
(iii) The use of chemicals is too expensive.
(iv) The use of chemicals is restricted by legislation. There is strong political pressure in Canada, for example, to ban the use of all synthetic chemical pesticides in its forests. This would leave products based on *B.thuringiensis* as the only insecticides registered for use in Canadian forestry.
(v) The markets are too small to support a new chemical development program. This applies to the horticultural sector, which has the added advantage, as far as microbials go, that many of the crops are grown in controlled environments. This should help with the problems of reliability mentioned earlier.

Conclusions

Under the right circumstances, microorganisms can be used as effective and reliable CP & E agents which offer commercially viable business opportunities. At present, these opportunities are restricted to markets which are small in relation to the markets for chemical CP & E agents. Despite its past reservations (many of which still persist) about microbial CP & E, the agrochemical industry currently has a more sympathetic attitude towards the concept than it has done for many years. This change in attitude is in part linked to the technological advances which now offer the opportunity for really effective and reliable products, which have generally been lacking in the past. There is a real desire to find good microbial CP & E products to complement chemicals already on the market. However, we should be realistic about what we can hope to achieve. Predictions by business analysts caught up in the wave of euphoria about biotechnology in the early 1980s, that microbials would have parity with chemicals by the turn of the century (Klausner, 1984) are grossly over-optimistic. Instead we should be looking for a gradual, modest increase in the contribution from microbials, accepting that chemicals will still be responsible for the major share of the market for the foreseeable future. However, industry's patience is limited and we need a real breakthrough in the next 3–5 years to give the subject real credibility. A major concern is that the chances of getting this breakthrough are severely jeopardized by a lack of fundamental knowledge in the area of microbial ecology—not in the much-publicized area of how to follow the fate of an organism introduced into the environment, although this is obviously important, but in the basics of how to identify the best organism for the job. It is simple to isolate microorganisms antagonistic to pests, but we know nothing about the characteristics required by an organism to integrate, survive and be active in the desired manner, when introduced into complex environments like soil and root or leaf surfaces. All the sophisticated molecular biology in the world will not help us if we do not increase our understanding in this area. Without the medium-term breakthrough, the industry's current level of interest in the subject will decline, and with it the chances of microbial CP & E making more than a token contribution to the global CP & E business as it does at present.

Acknowledgements

My thanks to Mr P.Davies, Elson's Seeds, Spalding, Lincs, Mr M.P.Greaves, Long Ashton Research Station, Drs J.R.Jebb and S.Potter, Agricultural Genetics Company, Cambridge, Dr N.J.Poole, ICI, Jealott's Hill Research Station, Dr P.Quilt, Biotal Ltd, Cardiff, Dr G.E.Templeton, University of Arkansas, and numerous colleagues in Shell companies for helpful discussions during the preparation of this chapter. Agrow, Biotechnology Affiliates and Wood-Mackenzie are gratefully acknowledged for permission to quote material from their publications.

References

Ager,B. (1987) Regulations/guidelines: developments in the UK. In *The World Biotech Report, Volume 1, Part 1. The Business of Biotechnology*. Online Publications, London, pp. 37–42.

Agrow (1985) *The World Agrochemical Market. Agrow World Agrochemical News*. George Street Publications Ltd, Richmond, Surrey.

Andrews,R.E., Faust,R.M., Wabiko,H. and Raymond,K.C. (1987) The biotechnology of *Bacillus thuringiensis*. *Critical Reviews in Biotechnology*, **6**, 163–232.

Biotechnology Affiliates (1987) *Biopesticides: Marketing, Technical and Legal Perspectives*. Biotechnology Affiliates, Reading RG8 0BP, UK.

Bowers,R.C. (1986) Commercialisation of Collego—an industrialist's view. *Weed Science*, **34**, (Suppl. 1), 24–25.

Burges,H.D. (1986) Production and use of pathogens to control insect pests. *Journal of Applied Bacteriology, Symposium Supplement*. 127S–137S.

Carsel,R.F. and Smith,C.N. (1987) Impact of pesticides on groundwater contamination. In Marco,G.J., Hollingworth,R.M. and Durham,W. (eds), *Silent Spring Revisited*. American Chemical Society, Washington DC, pp. 71–83.

Charudattan,R. (1985) The use of natural and genetically altered strains of pathogens for weed control. In Hoy,M.A. and Herzog,D.C. (eds), *Biological Control in Agricultural IPM Systems*. Academic Press, London, pp. 347–372.

Cook,R.J. and Baker,K.F. (1983) *The Nature and Practice of Biological Control of Plant Pathogens*. The American Phytopathological Society, St Paul, Minnesota.

Crespi,R.S. (1986) Patent issues in biotechnology. In Day,P.R. (ed.), *BCPC Monograph No. 34, Biotechnology and Crop Improvement and Protection*. BCPC Publications, Thornton Heath, UK, pp. 209–218.

Crull,A. (1985) Outlook for biotechnology in crop protection chemicals: challenge or opportunity. In *Biotechnology in Plant and Animal Agriculture*. Business Communications Company Inc., Stamford, Connecticut, pp. 42–48.
Environmental Protection Agency (1980) Guidelines for registering pesticides in the U.S., Subpart M, Data requirements for biorational pesticides, preamble and guidelines (draft). NTIS, Springfield, Vancouver.
Hacking,A.J. (1986) *Economic Aspects of Biotechnology*. Cambridge University Press, Cambridge.
Hassall,K.A. (1982) *The Chemistry of Pesticides*. Macmillan, London.
Jutsum,A.R. (1987) Commercial application of biological control: status and future prospects. In *Biological Control of Pests, Pathogens and Weeds: Developments and Prospects*. The Royal Society, London.
Kenney,D.S. (1986) Devine—the way it was developed—an industralist's view. *Weed Science,* **34**, (Suppl. 1), 15–16.
Klausner,A. (1984) Microbial insect control—using bugs to kill bugs. *Biotechnology,* **2**, 408–419.
Lewis,J.A. and Papavisas,G.C. (1987) Application of *Trichoderma* and *Gliocladium* in alginate pellets for control of *Rhizoctonia* damping off. *Plant Pathology,* **36**, 438–446.
Lysansky,S.G. (1984) Biological alternatives to chemical pesticides. In *Proceedings World Biotech Report 1984 (Europe), Vol. 1*. Online Publications, London, pp. 1455–1466.
McGaughey,W.H. (1985) Insect resistance to the biological insecticide *Bacillus thuringiensis*. *Science,* **229**, 193–195.
Markle,G.M. (1983) Registration of biorationals. In Allen,G.E. and Nelson,M.R. (eds), *Proceedings of the National Interdisciplinary Biological Control Conference*. USDA, Washington DC, p. 106.
Poole,N.J. (1987) Release of genetically engineered organisms. In *The World Biotech Report 1987, Vol. 1, Part 1. The Business of Biotechnology*. Online Publications, London, pp. 63–65.
Rowe,G.E. and Margartis,A. (1987) Bioprocess developments in the production of bioinsecticides by *Bacillus thuringiensis*. *Critical Reviews in Biotechnology,* **6**, 87–127.
Smith,R.J. (1986) Biological control of northern jointvetch (*Aeschynomene virginica*) in rice (*Oryza sativa*) and soybeans (*Glycine max*)—a researcher's view. *Weed Science,* **34**, (Suppl. 1), 17–23.
Templeton,G.E. (1986) Mycoherbicide research at the University of Arkansas—past, present and future. *Weed Science,* **34** (Suppl. 1), 35–37.
Templeton,G.E. and Greaves,M.P. (1984) Biological control of weeds with fungal pathogens. *Tropical Pest Management,* **30**, 333–338.
Templeton,G.E. and TeBeest,D.O. (1979) Biological weed control with mycoherbicides. *Annual Review of Phytopathology,* **17**, 301–310.
Van den Bosch,R. (1978) *The Pesticide Conspiracy*. Prism Press, Dorchester.
Wood-Mackenzie (1987) *Agrochemical Service*. Published by Wood-Mackenzie, Edinburgh EH2 4NS, UK.

CHAPTER 3

Global importance of *Rhizobium* as an inoculant

ALLAN R.J.EAGLESHAM

Boyce Thompson Institute for Plant Research, Cornell University, Ithaca, NY 14853, USA

Introduction

Although the restorative effects of legumes on agricultural soils have been recognized since antiquity, until a century ago root nodules were thought to be storage organs or of pathological origin. At a meeting of the Royal Society almost exactly 100 years ago (May 17, 1888), J.B.Lawes, the founder of Rothamsted Experimental Station, and J.H.Gilbert, his colleague of many years, presented a paper entitled 'On the present position of the question of the sources of nitrogen of vegetation'. They described to the British scientific establishment the landmark work of H.Hellriegel and H.Wilfarth which had been reported a few months earlier at a meeting in Germany.

> The new experiments were made with oats, buckwheat, rape, peas, serradella and lupins. The experimental soil was a pure sand, entirely free of nitrogen . . . to which were added the necessary mineral constituents. All the plants grew until the nitrogen of the seed was used up. Then to each pot a small quantity of the turbid watery extract of a surface soil was added . . . containing 0.3 to 0.7 mg nitrogen. . . . Neither the oats, rape nor buckwheat showed any effect from the addition of the soil extract, but remained in the condition of 'nitrogen hunger'. On the other hand, the Papilionaceae after a time recovered from their nitrogen-hunger, suddenly became dark green, and then grew luxuriantly . . . Thus, it may be considered established that the Papilionaceae can take the whole of their nitrogen from the air.

Lawes and Gilbert concluded:

> It seems natural to attribute the action to bacteria, and to connect it with the organisms in the nodules.

Beijerinck (1888) drew the same conclusion, and within months isolated nodule bacteria (designated *Bacillus radicicola*), and satisfied Koch's postulates. In 1895, the German scientists F.Nobbe and L.Hiltner, showing latter-day entrepreneurial acumen, filed patents for the process of inoculating seed with pure cultures of rhizobia. The inoculants were marketed under the name 'Nitragin' and at first were gelatin-based cultures (to be suspended in water); later, jelly, sterile soil and, in the 1920s, peat were developed as carriers (Fred *et al.*, 1932). The impetus for the development of the inoculant industry was the desire to introduce leguminous species to areas where they had not been grown before, with immediate success.

Prior to the mid-1970s there was relatively little research interest in nitrogen fixation by legumes, and activity was restricted to a few groups of scientists in Europe, the USA and Australia. The first meeting in the series which became the North American

Rhizobium Conference was held around a table at Iowa State University in 1968; but 15 years later at the Boyce Thompson Institute there were over 200 participants. The energy crisis of 1973–74 caused sudden, rapid increases in the costs of nitrogenous fertilizers and stimulated interest in legume nitrogen fixation and the possibilities of replacing chemical fertilizers with organic alternatives. Many less developed countries were particularly adversely affected by the increased costs of crop production, stimulating research on tropical legumes, even in developed countries.

In the early 1970s, the discovery that endonucleases could be used to clone fragments of DNA quietly heralded a new era in genetics research. The novel techniques were quickly applied to nitrogen-fixing organisms and the past decade has seen astonishing progress in our understanding of rhizobial genetics. We are now in the biotechnology age, and it is interesting to see that the new potentials for manipulating microbe−plant interactions have led to renewed interest in the old biotechnology of applying inoculants to legumes: major industrial funding has been invested in the USA to make a 'better' inoculant, and new companies have started operations in Europe and Canada. In 1988, a manipulated strain of *Rhizobium meliloti*, with multiple copies of nitrogen-fixing genes, will be field-tested with lucerne in the USA.

Appraisal of the impact of rhizobial inoculants on legume crop productivity is timely, and a global overview rather than a thorough and comprehensive review is presented in this chapter. A questionnaire was sent to 70 countries, to scientists with experience with rhizobia and legumes, in order to obtain current information on legumes grown and their importance, and on inoculant availability and usage. In the hope of encouraging scientists to impart information, the questionnaire was restricted to a single page, with invitation to comment on additional sheets (*Table 1*). Therefore, the potential for accumulation of detailed information was compromised by the objective of obtaining a general picture.

The information is presented below in ten sections representing geographical and/or climatic zones. Countries from which responses were received are listed for each zone, as are grain/vegetable legumes and pasture/forage/green manure/fibre legumes of importance.

Table 1. Questionnaire on legume cultivation, inoculant availability and usage.

Name
Institution, country or location of expertise
Legumes grown there
Which are of major importance?
Which are 'traditional' and which are recently introduced?
In your experience or opinion (delete as appropriate) on which of these legumes will inoculant application usually increase yield?
In cases where inoculation does not increase yield—why not?
Are inoculants available to growers? [If 'yes' proceed to (a), if 'no' proceed to (b)]
(a) Of what type are they (peat-based) and where do they originate?
What fraction of growers use inoculants?
On what legumes?
(b) Is there a need to make inoculants available to growers? If so, for what legumes?
Is peat locally available?
Other comments?

Legume inoculants: a global overview

Africa (West, East, Central and South)

Responses received from: Cameroon, Ghana, Kenya, Madagascar, Malawi, Mozambique, Nigeria, Rwanda, Sierra Leone, South Africa, Tanzania, Zaire, Zambia, Zimbabwe.

Grain/vegetable legumes: Cowpea, peanut, pigeonpea, bean (*Phaseolus vulgaris*), pea, faba bean, bambara, soyabean, lima bean, winged bean.

Pasture/forage legumes: Leucaena, clover, lucerne, medics, lupin, vetch, serradella.

Food legumes are of predominant importance, with the notable exception of South Africa (of which the south coast has a Mediterranean climate) where forages are also grown. The most important legumes are bambara groundnut (west), cowpea and peanut (all areas), and bean (east and south). The growing interest in soyabean noted by Ayanaba (1977) continues. Locally produced inoculants are available in Kenya, Madagascar, Malawi, Rwanda, South Africa, Tanzania, Zaire, Zambia and Zimbabwe. However, most African countries are without peat deposits, therefore only in Madagascar, Rwanda, South Africa and Zambia are inoculants peat-based. By-products of the sugar-cane industry are used as inexpensive carrier sources: filter mud in Kenya, Malawi and Tanzania, and bagasse in Zimbabwe. To a very large extent, inoculant use is with soyabean, grown for fodder, oil and export. Therefore the commercial producer, rather than the family farmer, benefits from the availability of inoculants. Except in a few localities, for example Tanzania and Nigeria, where it has been grown for many years, soyabean is a newly introduced crop, and inoculation is needed. Several correspondents reported the need for a greater public awareness of potential gains from inoculant use, for the family farmer; a good example is provided by Rwanda where high-quality soyabean inoculants are finding some acceptance at the small-farm level (Macary *et al.*, 1986). Possibilities of yield improvement of 'traditional' crops, such as bean in Kenya, Malawi and Tanzania, have been indicated by pilot trials.

Without exception, in the countries in which inoculants are not available there is a perception of need for them, if only to foster improved soyabean production. In several countries, scientists reported that traditional crops would respond to inoculation, for example cowpea in Burkina Faso and Nigeria, bean in Cameroon, bambara groundnut in Ghana, peanut in Mozambique. Of the countries in which inoculants are unavailable, only Mozambique has peat deposits.

Research at the *International Institute of Tropical Agriculture* (IITA) in Nigeria seeks to replace the traditional 'slash and burn' farming system with sustainable, continuously productive alternatives. Legume shrubs and trees (species of *Leucaena, Gliricidia, Flemingia, Cassia, Calliandra*) are being examined for 'alley cropping' sources of mulch and green manure (IITA, 1987).

The *African Association for Biological Nitrogen Fixation* held its first meeting in Nairobi in 1984, and the second in Cairo in 1987. The proceedings volume (Ssali and Keya, 1985) provides a recent review of current research objectives.

Australia and New Zealand

Grain/vegetable legumes: Pea, bean, lupin, chickpea, lentil, broad bean, pigeonpea, soyabean.

Pasture/forage legumes: Clovers, medics, lucerne, serradella.

All of the agricultural legumes grown in Australasia are of exotic origin, and inoculation is necessary in soils where they are being grown for the first time (however, see below). Peat-based inoculants were introduced in the 1940s. The voluntary but rigorous quality-control systems which were set up in Australia and New Zealand in the mid 1950s not only fostered acceptance by farmers, but also boosted scientific research.

Legume inoculation became an integral part of the ley farming system in Australia, thus providing nitrogen both directly for better pasture quality and indirectly for subsequent cereal production. Australia provides an excellent example of the potential significance of the high-quality inoculant as a tool for the improvement of agricultural productivity.

The widespread use of some of the forage legumes has resulted in their naturalization in Australia. This is particularly the case with clovers, and to a lesser extent with medics and lupins. Therefore, inoculation of these legumes may not be required, even in soils where they have not been planted before. Nevertheless, inoculant application is the advised practice.

The development of large areas of hill-country pasture refocused attention on legume inoculation in New Zealand in the 1970s. The major emphasis in New Zealand is firmly with forage and pasture legumes, the food legumes were reported for Australia only.

Despite the successful introduction and use of inoculants in Australia, some scientists believe there is potential for greater efficacy. Application to seeds at rates higher than recommended have been found to improve legume productivity when grown in a soil for the first time. An alternative strategy would be to improve the ability of the inoculant rhizobia to survive on the seed and to colonize the emerging radicle more efficiently.

Caribbean and Central America

Responses received from: Antigua, Costa Rica, Haiti, Jamaica, Mexico, Nicaragua, Panama, Puerto Rico, St Lucia, Trinidad and Tobago.

Grain/vegetables legumes: Bean, soyabean, pigeonpea, peanut, cowpea, lentil, Mexican yam, mungbean, winged bean.

Pasture/forage legumes: Siratro, leucaena, centro, pueraria, lucerne.

As in much of Africa, there is a predominant importance of grain legumes over forage. Soyabean is a recent introduction, interest in leucaena is mainly at the research stage.

Peat-based inoculants are produced in Costa Rica, Mexico and Nicaragua and used chiefly on soyabean. Although some of the more traditional crops, bean, chickpea and peanut, are inoculated in Mexico and Costa Rica, it is only to a small extent. Again, as in Africa, inoculant use has its chief impact at the commercial-grower and government-farm levels. In the countries where inoculants are unavailable there is a perception of need for them with traditional crops in Antigua, Haiti, Jamaica, and Panama; of these countries, inoculant-quality peat has been identified only in Jamaica. In Puerto Rico, St Lucia and Trinidad and Tobago, soyabean is not grown and nitrogen is not perceived as a limiting factor for traditional legumes, therefore inoculants are not required.

East Asia

Responses received from: People's Republic of China, Japan, Korea, Ryukyu Islands, Taiwan.

Grain/vegetable legumes: Soyabean, peanut, pea, mungbean, bean, faba bean, adzuki, lupin, lima, horse bean, cowpea.

Pasture/forage legumes: Astragalus, clover, vetch, lucerne, sesbania, leucaena, siratro, desmodium, trefoil, caragana, phasey bean.

Inoculants are used in all of these countries, and are locally produced except in the Ryuku Islands where Australian products are available. The inoculants are peat-based, except in a few localities in China where the liquid form is used, and in Korea where perlite is used as carrier.

Inoculants were introduced to China in the 1950s, and their importance varies from area to area. They are used most prominently with soyabean, peanut, mungbean, astragalus, clover and vetch. Since soyabean has its centre of origin in this part of the world, it is surprising to see that it is inoculated; yield increases are modest, however, in the range of 10%. The chief use of inoculants in the south and the Yangtze valley is with astragalus, for green manure production in rotation with rice. Lucerne acreage is being expanded in semi-arid areas of northwest China. Provincial agricultural colleges and institutes are responsible for inoculant production. Some scientists expressed concern about variable inoculant quality and use of inappropriate strains.

In Japan, modest seed-yield responses from inoculation have been reported with soyabean. White, red and ladino clovers and lucerne are inoculated in the more temperate, northern areas. However, inoculation is not a common practice among Japanese farmers; appropriate strains of rhizobia are indigenous, and nitrogenous fertilizers are relatively inexpensive.

Inoculant technology is new to Korea, and locally mined perlite is being developed as the base. Soyabean, pea and lucerne are the legumes of chief importance but short-term plans are to provide inoculants for soyabean only. Research trials indicate seed yield increases in the order of 10%.

The forage legumes, siratro, desmodium and leucaena are most important on Okinawa in the Ryukyu Islands. Commercial producers are the chief users of imported Australian inoculants, with recently introduced tropical forage production systems.

Multiple inputs characterize agriculture in Taiwan where soyabean, mungbean, peanut and adzuki are the important legumes. Inoculants are used only with soyabean and only to a small extent; indigenous rhizobia and high levels of residual nitrogen in soil commonly mitigate the need for inoculation. Research effort is being applied to using fast-growing legume trees for timber and pulp production; the role of inoculation is uncertain.

A.R.J. Eaglesham

Europe

Responses received from: British Isles, Czechoslovakia, Denmark, France, Italy, Romania, Soviet Union, Spain, Switzerland, Yugoslavia.

Grain/vegetable legumes: Pea, bean, faba bean, coccineus bean, lupin, soyabean, chickpea, lentil.

Pasture/forage legumes: Clovers, lucerne, sainfoin, vetches, hedysarum, trefoil.

In terms of total acreage, the legumes are not of major importance in Europe. Neither are they insignificant, however. Inoculants are available, although the frequency of use varies greatly from country to country; the appropriate rhizobia are usually indigenous and soil nitrogen levels are high. The most important legume is lucerne. *Rhizobium meliloti* is not native to European soils; however, frequency of inoculation varies, for example 100% of growers in Denmark apply inoculant, whereas only 6% do so in France. This may be a reflection of soil pH; long-term survival of lucerne rhizobia is poor at pH <6.5.

In the British Isles, field trials have indicated that bean and lupin respond to inoculant application in many, although not all, soils; if acreage of these increases, inoculant use would be advisable. Peas (for food and fodder) and faba beans (fodder) are the most important legumes, and neither requires inoculation. Three commercial inoculant-production companies have recently been established in England, the chief market being the soyabean industry in Europe, particularly Italy, and in Canada.

In Spain and Czechoslovakia, soyabean is being investigated as a new crop. In Spain, soyabean and hedysarum are the only legumes which require inoculation, and they are of minor importance in comparison with bean, chickpea, lentil, faba, lucerne and vetch; if soyabean increases in importance in the future, locally produced peat inoculants will be available. In contrast, peat inoculants in Czechoslovakia are imported from the Soviet Union; a relatively high fraction, 25%, of growers of clover and lucerne apply inoculants. Inoculant use is similarly common in the Soviet Union; again, acidity may contribute to poor survival of rhizobia in podzolic soils.

Planting of fodder pea has recently increased in Denmark and now is planted on 3% of total acreage; inoculation is not required.

Soyabean acreage has expanded rapidly in recent years in France, Italy, Romania and Yugoslavia, and it is now the grain legume of choice in the last three countries. The absence of *Bradyrhizobium japonicum* from European soils dictates inoculant application where the crop is first planted; however, it appears that inoculation has become a routine practice even where the crop has been planted previously.

Agriculture is a minor industry in Switzerland; legume acreage is modest and inoculants are not used.

North Africa and West Asia

Responses received from: Cyprus, Egypt, Ethiopia, Iraq, Israel, Jordan, Morocco, Oman, Somalia, Sudan, Syria, Tunisia, Turkey.

Grain/vegetable legumes: Faba bean, chickpea, lentil, pea, bean, peanut, pea, lupin, cowpea, soyabean.

Pasture/forage legumes: Clovers, medics, lucerne, vetches, siratro, hedysarum, fenugreek, berseem.

Food legumes, particularly chickpea, faba bean and lentil, rate with cereals as important foods in this part of the world. However, as far as inoculant usage is concerned, there is little to report except for Egypt and Israel. With no peat deposits in Egypt, Nile silt and charcoal have been used since the mid 1950s to manufacture inoculants. The construction of the Aswan Dam eventually brought problems of supply of silt, therefore imported peat is now used. Alternative locally available carriers are under review, and cottonseed-husk compost has recently been found promising. Soyabean was recently introduced to Egypt and it currently constitutes 5% of legume acreage. It is not surprising that inoculation is essential for this newly introduced crop, but several traditional winter crops routinely respond to inoculation with yield increases in the range 20−30%: clover, faba bean, lucerne and lentil. The reason for this yearly response is unclear, but elevated soil temperatures coupled with desiccation, during summer months, may result in depletion of rhizobial numbers. In Israel, the majority of peanut growers routinely apply inoculants and positive responses to inoculation are obtained year after year, again presumably because of poor survival of introduced rhizobia between crops.

Of the other countries in this group, inoculants are available in Morocco, Oman, Tunisia and Turkey. Introduction of soyabean to Morocco in the early 1980s as an oil crop led to importation of inoculants; with imported peat, local production was initiated in 1987. Expected increase in chickpea acreage, as a result of adoption of winter planting, is expected to rationalize production of inoculant also for this crop. In the Sultanate of Oman, imported peat-based inoculants are used in government-sponsored projects and on large-scale farms, for pasture improvement. Locally produced inoculants are used with hedysarum in Tunisia, whereas in Turkey inoculants are produced predominantly for soyabean.

In Cyprus, Ethiopia, Iraq, Jordan, Somalia and Sudan the perception is of a pressing need for inoculants, not only for soyabean introduction, but also for traditional crops grown by the family farmer. Field data gathered at the *International Centre for Agricultural Research in Dry Areas (ICARDA)*, in Syria, has demonstrated the potential for an important role for inoculants in the region. Although lucerne and medics originated in this part of the world, they have long been treated as weeds on arable land, resulting in depletion of rhizobial numbers in many soils; thus striking responses to inoculation are now common. Replacement of the fallow-farming system by a legume-based ley system would have enormous impact on agricultural productivity. Current work at ICARDA indicates that the food legumes, chickpea and lentil, will also respond strongly to inoculation in many soils. Peat availability is one hurdle to be addressed; deposits of good quality exist in Turkey and could be exploited for the region. Alternative carriers are being researched.

The current status of research in this area was recently reviewed (Beck and Materon, 1988).

A.R.J. Eaglesham

North America

Responses received from: Canada, United States of America.

Grain/vegetable legumes: Soyabean, bean, peanut, faba bean, pea, lentil, cowpea, chickpea, lima, lupin.

Pasture/forage legumes: Lucerne, clover, trefoil, sainfoin, melilotus, indigofera, aeschynomene.

All of the agricultural legumes of major importance in North America are exotic in origin and have been introduced relatively recently (mainly in the last 100 years). Only species of *Indigofera* and *Aeschynomene* are indigenous.

Surveys of farmers in Canada have shown a general awareness of inoculants. Indications are that more than 60% of legume crops are inoculated. If the time-lapse between plantings of grain legumes (pea, faba bean, bean, lentil, chickpea, soyabean) is 4 or more years, re-inoculation is recommended. Forage (clovers, trefoil, lucerne, melilotus) fields are usually planted in rotation after several years of annual cereals; therefore the general recommendation is to inoculate all plantings. Up to present, the inoculants available in Canada originated in the USA, and to a small extent in the UK and Australia. For the coming planting season, however, it is expected that there will be 3 new inoculants companies in Canada. Over the past 2 years there has been a shortfall in availability of inoculants for pea and lentil.

Although soyabean had been grown in the USA since the mid-1700s it was not until the early years of this century, a period of rapid agricultural expansion, that significant acreage was planted. Originally, soyabean was grown for forage; the first oil and meal processing facility was established in Seattle in 1911. By that time it had been realized that the appropriate rhizobia for soyabean, and for other newly introduced legume crops, were absent from soils in the USA, and a number of brands of inoculants of various types were available (Fred *et al.*, 1932). Figures for 1980 show that total area inoculated was in excess of 15 Mha; 73% of inoculants produced in the USA were for soyabean, 9% each for lucerne and peanut, 5% for pea and the vetches and 4% for clovers (Burton, 1982). Certainly where a crop is being planted for the first time, inoculation is advisable, but the question of need to inoculate a crop which has become 'traditional' is often debated. Consequently, estimates indicate that 70% of soyabean growers in the Atlantic, southeastern and delta states of the USA, where soyabean acreage continues to increase, apply inoculants, whereas in the midwest probably less than 20% of soyabean is inoculated. On soils in which soyabean has been grown within the previous 4 years, positive responses to inoculation are unusual; yield increases of perhaps 10% may be expected 20% of the time. This frequency of response would justify routine use of inoculants on soyabean as an insurance policy. Edaphic factors, for example pH <6.0 and low organic matter, can adversely affect long-term rhizobial survival, and *may* further justify routine inoculation. Positive responses to inoculation are normally obtained with chickpea, lucerne and clover planted in acid soils of the northwest.

A range of types of inoculants, including liquid- and clay-based powder, are available in the USA, but the seed-applied peat-based type continues to dominate the market. Granular inoculants, available as clay- or peat-based, were developed chiefly for use with peanut to circumvent problems of seed-coat damage resulting from application of the powder type. Granular inoculants also offer the potential to avoid contact with seed-applied pesticides and to impregnate soil at very high inoculation rates, but their cost to the grower is significantly higher than that of seed-applied types. Over recent years there has been an increasing trend for seed companies to supply lucerne, and other small-seeded legumes, to the grower pre-inoculated with rhizobia, despite the fact that there is a great deal of evidence that long-term survival of rhizobia is poor on seeds, mainly as the result of desiccation. Recently we made a viable count of bacteria on locally-supplied pre-inoculated lucerne, and found only 17 cells per seed; tested isolates did nodulate (J.M.Ellis and A.R.J.Eaglesham, unpublished data). A national inoculants producer claims counts of 10^3 per pre-inoculated lucerne seed for at least 5 months; another producer quotes no numbers per seed but states an average $\log_{10}$-reduction time of at least 6 months with a 1-year expiry date. A good quality sterile-peat inoculant will deliver 5×10^3 cells per lucerne seed.

There is concern that potential gains from inoculation are being lost. Many farmers 'dry inoculate' by mixing powder inoculant directly with seeds in the planter, with the result of poor and uneven adhesion. Manufacturers have changed formulations to improve adhesion with this method, but clearly slurry-preparation with an adhesive is still advisable. Another problem is contact with seed-applied fungicide. Granular inoculants provide an answer, but many farmers would be reluctant to make the larger financial investment in equipment and higher inoculant cost. There is no federal prescription for inoculant quality in the USA, and only 4 states have minimum requirements. These standards are defined in terms of presence or absence of nodules in a greenhouse test, which could be met by inoculants with very low counts and are therefore meaningless. A survey of inoculants available in the southeast demonstrated variable quality (Skipper *et al.*, 1980).

South America

Responses received from: Argentina, Bolivia, Brazil, Colombia, Guyana, Peru, Uruguay, Venezuela.

Grain/vegetable legumes: Soyabean, bean, pea, broad bean, peanut, pigeonpea, chickpea, cowpea.

Pasture/forage legumes: Lucerne, clover, melilotus, lotus, lupin, crotalaria, glycine, mucuna, stylo, pueraria, siratro, centro, gliricidia.

Bean is a daily constituent of the diet throughout Latin America, whereas soyabean is now widely grown for fodder, oil and export. Livestock production is important in many countries of the region, therefore forage and pasture legumes are also of agricultural significance.

Peat-based inoculants are produced in Argentina, largely for use with soyabean but also with lucerne, trefoil and melilotus. Soyabean was introduced in the early 1960s, and its rapid expansion has made Argentina the fourth largest producer (3 Mha, after the USA, Brazil, and the People's Republic of China), with inoculant applied to at least 60% of plantings.

The common bean is less important in Bolivia than in other countries of the region, its place taken by faba bean and pea. Soyabean acreage is rapidly expanding in Bolivia, with inoculants imported from Brazil and the USA. There is hope that the pressing need for locally produced soyabean-inoculants will be resolved with foreign aid in the near future.

In Brazil, soyabean production has increased 25-fold in the last 23 years and the area planted is approaching 50% of that in the USA. There are several producers of peat-based inoculants, which are used on at least 90% of the Brazilian crop. Therefore, total soyabean areas routinely inoculated are now approximately equal in Brazil and the USA (8 Mha). Forage and pasture legumes, for example lucerne, clovers and lupins in the south, and stylo, centro and siratro in the north, and green manures, such as crotalaria, mucuna and pueraria, are inoculated perhaps as much as 30%. However, inoculant use with the food legumes important to rural communities, bean and cowpea, is insignificant. Tree legumes are under study in Brazil (by Empresa Brasileira de Pesquisa Agropecuaria) for fuel, timber and pulp, and the role of inoculation is a component of that work (EMBRAPA, 1984).

Despite the availability in Colombia of peat-based inoculants produced locally as well as imported from the USA and France, their use has not become an accepted farm practice. Some large-scale producers inoculate soyabean and pasture legumes, clover, lucerne, pueraria and stylo, but not so at the small-farm level. Work at the *Centre for Tropical Agricultural Research (CIAT)* near Cali has demonstrated that inoculation is an essential component for the improvement of pasture productivity of the impoverished acid soils of the Amazon Basin (Sylvester-Bradley *et al.*, 1988). How the technology will be transferred remains to be decided.

The situation in Guyana and Peru appears to be similar to that in Colombia as inoculants are produced but little used. No doubt the cycle of limited resources and meagre productivity are relevant factors. In contrast, inoculants are widely used in Uruguay, having been introduced in 1960 for forages, clovers, lucerne, medics and trefoil, and expanded in 1973 for soyabean. An important component is the national system of quality control, with checks made at production sources and also at retail centres. As has occurred in Australia and New Zealand, quality control has fostered both research and farmer-confidence. In fact research continues in Uruguay, with the objectives of expanding inoculant use to include pea, lentil and chickpea, and for better understanding of environmental stresses on the establishment and persistence of legume nitrogen fixation.

Inoculant production was recently initiated in Venezuela; in 1982, 90 ha were inoculated, growing to 3000 ha by 1986 and projected to increase to 150 000 ha in 1990. The introduction of soyabean provided the impetus for inoculant production, and it is interesting to note a 'spin-off' effect, that some growers are using inoculants also on cowpea and bean.

The proceedings volume from a workshop on inoculants in Latin America was recently published (Freire, 1987).

South Central Asia

Responses received from: Bangladesh, Bhutan, India, Nepal, Pakistan, Sri Lanka.

Grain/vegetable legumes: Lentil, chickpea, pigeonpea, peanut, pea, cowpea, lathyrus, mungbean, black gram, soyabean, lablab, horsegram, broad bean, lima, winged bean.

Pasture/forage legumes: Berseem, lucerne, sesbania, crotalaria, clover.

Food legumes are of prime importance for human consumption in this part of the world where vegetarianism is a common philosophy. Only soyabean, berseem and white clover are recent introductions.

In Bangladesh, on-farm trials over the last three years in agriculturally important areas have indicated accumulation of significant benefits (yield increases in the range 20–90%) from inoculation, with all of the important food legumes (lathyrus, lentil, chickpea, mungbean, cowpea, black gram, peanut and soyabean). Inoculants are now being made available to Bangladeshi farmers for the first time using locally mined peat. Lablab, pea and cowpea are grown for green vegetables, crotalaria for fibre (sunn hemp) and sesbania as a green manure.

White clover has very recently been introduced to Bhutan in a move to improve pasture quality. Inoculation is required and a locally produced peat-based type is available.

During the introduction of soyabean to India on a large scale in the mid-1960s, the necessity to inoculate was realized, with concomitant spectacular increases in yields. In the 1970s, the increasing costs of nitrogenous fertilizers encouraged extrapolation from the soyabean experience, and much industrial and research effort went into production of inoculants for other legume crops. The technology was taken to farmers in haste, some products being of poor quality and with no determination of the need to inoculate. As a result, inoculants enjoyed a short-lived popularity, and farmers became sceptical of their utility; today perhaps as few as 5% of legume plantings are inoculated. A chief reason for lack of response to inoculation is that the most productive soils are used for cereal production and to a large extent legumes are planted on marginal lands on which occur deficiencies of nutrients other than nitrogen. Much of India is arid, and responses to inoculation are unlikely under moisture-limited conditions; again, irrigated areas are likely to be used for cereal production. Many farmers are reluctant to apply inputs of any kind to their legume crops. Inoculants are made available to farmers via regional agricultural universities and private companies. Currently a 'national biofertilizer centre' is being established, with subcentres to be located throughout the country. Most of the inoculants produced are charcoal- or lignite-based, some in combination with organic soil; peat-based inoculants are produced in Kerala State. Chickpea and pigeonpea are the major food legumes for India, followed in importance by lentil, pea and cowpea; peanut and soyabean are grown as oil-seed crops. Berseem is an important forage in the Punjab, and is frequently inoculated. Several respondents expressed the belief that inoculants would have a beneficial effect on traditional legumes, provided they were grown under fertile conditions.

In Nepal, soyabean, lentil, cowpea and faba bean are of chief importance and field trials have indicated strong responses to inoculation. These trials were done with isolates from local soils. Inoculant trials are planned for other areas of Nepal. At present inoculants are not available to farmers.

Forage legumes are of greater importance in Pakistan than in India, although a broad range of food legumes is also produced. A pilot plant to provide soil-based inoculants for research purposes has recently been set up at the *National Agriculture Research*

Centre in Islamabad. The perception is that all of the important legumes, with the possible exception of chickpea, will respond to inoculation: lentil, mungbean, black gram, peanut, cowpea, faba bean, soyabean, vetches, siratro and berseem.

Soyabean was recently introduced to Sri Lanka and inoculants are imported from the USA. Cowpea (particularly as long bean), black gram and mungbean are of major importance as food legumes, and crotalaria and sesbania are used as green manure for rice. As in Pakistan, there is a belief that most of the traditional legume crops would show improved productivity if inoculants were available.

Southeast Asia

Responses received from: Indonesia, Malaysia, Philippines, Thailand, Vietnam.

Grain/vegetable legumes: Soyabean, peanut, mungbean, pigeonpea, cowpea, long bean, rice bean, bean, pea, winged bean.

Pasture/forage legumes: Centro, calopogonium, mucuna, pueraria, leucaena, stylo, sesbania, gliricidia.

Much of the legume/*Rhizobium* work in Indonesia and Malaysia has been ancillary to research on rubber and oil palm, mainly in terms of maximizing inputs of nitrogen from cover crops during the years of grove-establishment prior to canopy closure. Scientists at the *Rubber Research Institute* in Malaysia have developed a cover-crop succession system. Seeds of *Mucuna cochinchinensis, Pueraria phaseoloides* and *Calopogonium caeruleum* are planted simultaneously; mucuna establishes rapidly, thus suppressing weeds, but is, in turn, suppressed by pueraria within several months, and caeruleum eventually dominates when shading by the rubber becomes a factor. The smallholder, however, sees little tangible benefit from the expense and effort of planting the cover crop. Therefore, another research objective is to devise methods of land use in the newly planted tree groves for immediate, as well as long-term gain, for example by planting peanut. Inoculant use for cover-legumes, therefore, is common on the plantation, but not on the family farm. Malaysian inoculants are made with a soil and coir dust mixture as carrier, and in Indonesia they are peat-based.

Soil-based inoculants are produced in the Philippines but have made little impact. The important food legumes are mungbean, cowpea and bean, and field trials have shown minimal responses obtained erratically, probably due to the presence of effective indigenous rhizobia. At the farm level it would seem likely that low yield potential would further militate against positive effects from inoculation.

In Thailand about 100 tons of peat-based inoculant are produced per year in Bangkok, mainly for application to soyabean which has been found to respond strongly in soil where it has not been previously grown—about 10% of the crop. To a very much smaller extent, peanut and mungbean are also inoculated. In the Chiangmai area to the north, inoculants are not available, but required for soyabean.

Production of peat-based inoculants was initiated in South Vietnam for use with an expanding soyabean cultivation, but decline in world soyabean prices has led to a de-emphasis with the current result of little interest in inoculation.

Conclusions from overview

Inoculants have made an enormous impact on global agriculture particularly in the last 20 years in association with soyabean in the USA, Brazil and Argentina. Soyabean aside, significant contributions to agricultural productivity have accrued from the use of inoculants in Australia, North America, Eastern Europe, Egypt, Israel, South Africa, New Zealand and, to a lesser extent, Southeast Asia. However, as for the large majority of less developed countries of Southeast Asia, the Indian subcontinent, West Asia and Africa, and Central and South America, inoculant technology has made essentially no impact on productivity of the family farm.

The future

It is ironic that inoculants have had so little effect on legume production (again, soyabean aside) in the less developed countries in which they are available, whereas in those countries in which inoculants are unavailable there is perception of need for them. Yields of food legumes in farmers' fields are often meagre, and fall far short of those obtained under research-station conditions (Summerfield and Lawn, 1987). Unfortunately, too often it has been assumed that nitrogen is the chief yield-limiting factor, and inoculation touted as a panacea, thus rationalizing elaborate rhizobial-strain selection programs and field inoculant trails. Whether or not responses to inoculation can be demonstrated under fertile, high-yielding conditions is actually of little significance to the farmer growing legumes with no inputs and much lower yield potential. Although we should not rule out the possibility for inoculant benefit in improving a meagre yield to a poor one, the pressing need is to better understand the factors responsible for those meagre yields and how they may be assuaged. Assuming that higher yield can be achieved (Summerfield and Lawn, 1987), the need for inoculant, as a component of the 'package' of inputs, becomes an issue. As an adjunct to this long-term approach, there is a short-term need for information obtained by growing legumes exactly as farmers do. On-farm evaluation of inoculant technology will provide not only realistic information on potential impact on yields, but also demonstrate what problems the farmer encounters in adopting the new practice. If trials are done at the research station rather than on the farm, then inoculant effects should be examined without ancillary inputs, to mimic farm conditions as closely as possible (Somasegaran and Hoben, 1985; Wynne *et al.*, 1987). It seems likely that in many of the countries where need for inoculant availability is perceived for traditional food legumes, in fact there is little or no potential for benefit to growers, *at present yield levels*.

In the foreseeable future, the greatest potential for inoculant impact on agricultural productivity appears to be via pasture legumes in two areas of the world: South America and West Asia/North Africa.

At the *International Centre for Tropical Agriculture (CIAT)* in Colombia, work in the Tropical Pastures Programme has demonstrated that inoculation is a key component for the improvement of the quality and productivity of grass/legume pastures over vast areas of impoverished acid soils of tropical America. The legumes (species of *Pueraria, Stylosanthes, Centrosema, Desmodium* and *Arachis*), which have been selected for tolerance of these soils, show significant improvements in yields in response to inoculation (Sylvester-Bradley, 1984, 1988). A network of collaborating scientists, a microbiologist and an agronomist from each of six countries, has been set up to evaluate responses to inoculation under local conditions, using *Rhizobium*—legume combinations provided by CIAT (Sylvester-Bradley *et al.*, 1988).

At a sister centre at CIAT, the *International Centre for Agricultural Research in Dry Areas (ICARDA)* in Syria, research has indicated that adaptation of the ley farming system used in Australia could have enormously beneficial effects on the agriculture of North Africa and West Asia (Cocks, 1988). The concept is to reduce the costs of annual resowing and harvesting of pasture crops by using medics able to regenerate naturally and which are harvested by grazing; a nitrogen benefit from the legume accrues to the following cereal, and the legume regenerates and reseeds in the season following the cereal. It is estimated that 30 Mha of fallow land are suitable for this system, sufficient to support 100 million sheep, with a nitrogen accretion to the soil of 1.4 Mt, 65% greater than the amount of fertilizer nitrogen currently applied to Algeria, Tunisia, Libya, Jordan, Turkey, Iran and Afghanistan combined (Cocks, 1988). Poor productivity of some of the medic accessions which were screened at ICARDA was later found to be due to incompatibility with local rhizobia; inoculant application resulted in huge yield increases. Thus, as in South America, inoculation is a key component in the improvement of agricultural productivity in West Asia and Northern Africa.

In many of the countries within the mandates of both CIAT and ICARDA there is no inoculant industry and no peat available. The problem of transfer of inoculant technology thus is exacerbated. A large number of solid materials have been investigated for efficacy as alternatives to peat as inoculant carrier (reviewed by Williams, 1984; Smith, 1987). A novel approach to inoculant production which circumvents solid carriers is the oil-based system which appears to be particularly efficacious under hot, dry conditions (Kremer and Peterson, 1982, 1983); lyophilized rhizobia are protected from moisture by suspension in oil. Although the sophistication of lyophilization may appear inappropriate for a less developed country, in fact the pharmaceutical industry carries out large bulk freeze-drying procedures which could readily be adapted for rhizobia. We have prepared oil-based inoculants using strains and seeds of interest at ICARDA and have obtained acceptable cell numbers per seed (*Table 2*); however, loss of viability after incorporation in oil would appear to dictate that the farmer would have to incorporate the lyophilized cells in the oil immediately prior to application to seeds. In a dry Syrian soil, rhizobial survival on seed was inferior with an oil-based inoculant compared with a peat-based type, however after 7 days there were in excess of 10^3 cells per seed (*Table 3*); nodulation and yield data from this study are still to be determined, with water application at planting or at 7 or 14 days. If solid carriers are to be developed, an important criterion for small-scale production is the ability to support growth of rhizobia, as distinct from maintaining high numbers added in broth. Injection of a 1000-fold dilution of broth in water into sterile peat produced rhizobial counts of 10^9 g^{-1} after a few days (Somasegaran and Halliday, 1982; Somasegaran, 1985); thus, unsophisticated fermenters of modest volume can be used to produce significant quantities of inoculant.

It was discovered in Tanzania that some lines of soyabean, of southeast Asian origin, nodulated with indigenous rhizobia, showed no yield response to inoculation (Chowdhury, 1977). This presented the possibility of crossing lines from the USA and

Table 2. Statistics for an oil-based inoculant (*Rhizobium meliloti* strain M38 in olive oil): area of lucerne which 1 ml will inoculate, at 5×10^3 cells per seed and 5×10^5 seeds ha^{-1}.

	Cells ml^{-1}	Area ml^{-1} (ha)	Peat equivalent[a] (g)
Freshly prepared[b]	3.8×10^{11}	150	380
After 7 days	1.3×10^{10}	5.2	13
After 14 days	5.8×10^9	2.3	5.8

[a]Weight of peat required to inoculate the same area, assuming 10^9 cells g^{-1}.
[b]From a lyophilized preparation of 5.1×10^{12} cells g^{-1}, suspended in oil at 1:10 w/v.
Data from J.M.Ellis and A.R.J.Eaglesham, unpubl.

Table 3. Comparison of oil- and peat-based inoculants: survival of *Rhizobium meliloti* strain M38 on seeds of *Medicago polymorpha* in an air-dried Syrian soil.

	Inoculant (cells per seed)	
	Oil	*Peat*
Freshly inoculated	1×10^5	3×10^5
After 2 days	1×10^4	2×10^5
After 7 days	2×10^3	5×10^3
After 12 days	0	2×10^3

Values are means of six replicates. Average temperature regime of 21/10°C and relative humidity of 41/80%, day/night.
Data from L.A.Materon, J.M.Ellis and A.R.J.Eaglesham, unpubl.

lines from Asia to produce high-yielding types which would nodulate with indigenous rhizobia and not require inoculation for maximum yields. The observation that Asian lines nodulated well also in Nigerian soils led to the belief that the rhizobia responsible were of the 'cowpea' miscellany and were widely distributed throughout Africa (Nangju, 1980). However, a comparison of 'promiscuous' soyabean and cowpea revealed that they are not nodulated by the same rhizobia (Eaglesham, 1985). Also, it is dangerous to make assumptions regarding the characteristics of rhizobia in a soil; the rhizobial populations at three West African locations were found to be distinct, and even to contain a previously undescribed type of *Bradyrhizobium* (Eaglesham *et al.*, 1987). A recent screening of 79 'promiscuous' lines of soyabean and 42 lines of cowpea grown simultaneously in the same field at Abakiliki in Nigeria revealed that, at 25 days, the large majority of the soyabean lines had no nodules, whereas all the cowpeas were nodulated. The maximum numbers of nodules per plant were 3.3 and 52 for soyabean and cowpea, respectively; the minimum number of nodules per cowpea was 12 (G.U.Okereke and A.R.J.Eaglesham, unpublished data). Until we have a thorough understanding of the ranges and patterns of diversity of rhizobia in a soil in which a legume is likely to be grown (a tall order indeed!), it would appear that a safer philosophy is to breed for specificity with inoculant rhizobia producing effective symbioses rather than to breed for promiscuity with strains of unknown potential.

Problems of desertification and of fuel-wood availability indicate that research on fast-growing trees, including legumes, will become increasingly important. A review of recent work on tropical leguminous trees is available (EMBRAPA, 1984).

Table 4. Comparison of bean (*Phaseolus vulgaris* 'BAT 76') and cowpea (*Vigna unguiculata* 'VITA-3') for nodulation, acetylene reduction activity (ARA) and growth after 35 days in the greenhouse (conditions as in Eaglesham *et al.*, 1983).

Legume	Nodules per plant	Nodule dry wt (mg plant)	ARA (µmol plant h)	Shoot dry wt (g plant)
Bean[a]	748 ± 66	470 ± 54	99 ± 15	8.44 ± 0.56
Cowpea[b]	140 ± 40	670 ± 136	106 ± 23	6.60 ± 0.45

[a]Beans supplied with 50 mg N (KNO_3) at planting and inoculated with *Rhizobium leguminosarum* bv. *phaseoli* strain CIAT 144.
[b]Cowpeas supplied with 60 mg N (KNO_3) at planting and inoculated with *Bradyrhizobium* sp. (*Vigna*) strain IRc 256.
Values are means ± SD. Data from B.D.Eardly and A.R.J.Eaglesham, unpubl.

Finally, a consideration of possible impact of genetic engineering on *Rhizobium*—legume inoculant technology is appropriate. With the present state of the art, only the microsymbiont is amenable to molecular genetic manipulation. Much has been written about making the nitrogen-fixing symbiosis more 'efficient'. However, the common observation that an effectively nodulated soyabean crop will show little or no response to application of fertilizer nitrogen (Mengel *et al.*, 1987) indicates either that growth on fertilizer also requires to be made more efficient, or that yield potential is realized when the symbiosis provides a significant source of nitrogen. When soyabean was 'fertilized' with carbon dioxide to provide a 50% increase in yield potential, there was a five-fold increase in nitrogen fixation without an increase in nodule weight per plant (Hardy and Havelka, 1976) showing that nodule function was not yield-limiting. At present, the best candidate as a gene to improve efficiency of the root nodule symbiosis would appear to be that for hydrogen recycling, *hup*. The theoretical 10—20% benefit in nitrogen fixation efficiency which accrues when a rhizobial strain can recycle lost hydrogen has been tantalizingly difficult to demonstrate unequivocally under field conditions (Eisbrenner and Evans, 1983). Progress in tranferring *hup* genes will allow comparisons of strains which are isogenic apart for *hup*, and thus a definitive answer may be obtained (Lambert *et al.*, 1985).

Common bean is often regarded as a poorly efficient fixer of nitrogen, because of the frequency with which it has been observed to respond in the field to fertilizer nitrogen. Under greenhouse conditions, however, we have found that cv BAT 76, regarded as a mediocre fixer at CIAT (J.Kipe-Nolt, pers. comm.), when effectively nodulated shows patterns of nodulation and acetylene reduction equal to those of cowpea (*Table 4*). Bred lines of cowpea were found to respond strongly to fertilizer nitrogen in soils in which lines used by farmers did not (Ahmad *et al.*, 1981). Again this illustrates a need for better understanding of the characteristics of indigenous rhizobia and their potential effectiveness with new or introduced legumes. There is also much scope for investigation of effects of environmental stresses, particularly in combinations, on the *Rhizobium*—legume interaction, which impinge on determinations of need to inoculate. For example there has been significant effort in documenting the effects of high temperature and of moisture stress individually, but nothing is known of their effects together (Eaglesham and Ayanaba, 1984).

The possibility of introducing genes to benefit the host legume by manipulating the infecting rhizobia is an interesting concept. Rhizobia are known to produce compounds, in addition to ammonia, which can pass into the plant and have a modifying effect. The best example is rhizobitoxine, an amino acid homologue of cystathionine of molecular weight 204, which is produced by strains of *Bradyrhizobium* and induces chlorosis in leaves of some cultivars of soyabean (La Favre and Eaglesham, 1986). Another example of a compound produced by rhizobia in the nodule, ancillary to nitrogen fixation, is one which causes a 'leaf-roll' symptom in pigeonpea (Kumar Rao *et al.*, 1984); we have a strain which produces similar symptoms in mungbean but not in pigeonpea (A.R.J.Eaglesham, unpublished data). This raises the possibility of genetically engineering rhizobia to induce the formation of compounds which would confer advantages, for example pesticide resistance or insect resistance, on the host. The nitrogen status of soyabean has been found to correlate with relative palatability to predator insects. Cabbage loopers developed faster, grew larger and survived better on nitrate-fed plants, whereas Mexican bean beetle developed faster, grew larger and survived better on nitrogen-fixing plants; preliminary evidence indicates that presence or absence of compounds ancillary to the symbiosis, rather than nitrogen nutrition *per se*, are causal (P.R.Hughes and A.R.J.Eaglesham, unpublished data).

A major problem in getting superior rhizobia (genetically engineered or otherwise) into nodules of field-grown legumes can rise from competition from infective but inefficient indigenous strains. As a general rule of thumb, Weaver and Frederick (1974) suggested that cell number per seed has to be 1000-fold higher than numbers per g of soil to form 50% of the nodules. Percent nodule occupancies by inoculant rhizobia are commonly in the single digits. Many factors other than numbers may influence relative competitiveness of rhizobia, but how they act and interact is poorly understood (Dowling and Broughton, 1986). The ability to survive on the seed coat, then to move from the seed coat to the emerging radicle, and then to remain in contact with the root as it moves through the soil, appear to be problematical for rhizobia (Salema *et al.*, 1982a,b; Madsen and Alexander, 1982); amelioration of these conditions could do much to improve nodule occupancy by inoculant rhizobia.

General conclusions

Inoculant technology has had little impact on food production at the small-farm level. The potential for positive effects in less developed countries is difficult to assess at the present low potentials for yield. With inputs to raise yield ceilings, inoculants may, eventually, have an optimizing role; the inoculant is not a panacea. Areas of the world where there is clear potential for a positive role for inoculants are (i) pasture production on acid soils of South America, and (ii) ley farming in West Asia and North Africa; in both regions problems of local production and utilization of inoculants must be addressed.

Before there is scope for the improvement of impact of rhizobial inoculants as a result of genetic engineering, many fundamental problems relating to rhizobial competitiveness, agronomy and biology must be addressed.

Acknowledgements

I am grateful to all who returned the questionnaire, particularly to those who completed more than one, who added comments, included reprints and reports etc., and to those who passed the questionnaire to colleagues.

Africa: R.C.Abaidoo, A.Adebayo, D.S.Amara, K.O.Awonaike, A.Ayanaba, S.K.A.Danso, D.S.Daramola, K.E.Dashiell, A.Hakizimana, S.O.Keya, D.J.Khonje, N.Loyindula, D.Montange, K.Mulongoy, M.P.Nuti, R.C.Nyemba, G.U.Okereke, M.R.Ryder, V.Ranga Rao, H.Saint Macary, M.P.Salema, P.Singleton, J.L.Staphorst, B.W.Strijdom.
Australia and New Zealand: P.M.Bonish, J.Brockwell, J.H.A.Butler, D.L.Chatel, D.F.Herridge, J.C.Howieson, J.Keoghan, W.L.Lowther, C.A.Parker, R.J.Roughley, J.A.Thompson.
Caribbean and Central America: M.H.Ahmad, B.Ahmed, A.Belen Vesga Cala, J.F.Felix, B.Hernandez, D.H.Hubbell, J.J.Pena Cabriales, C.Ramirez Martinez, H.Saint Macary, E.C.Schroder, M.Valdes.
East Asia: Ge Cheng, Ching-sen Fan, G.C.G.Fernandez, Ning Guo-zan, T.S.Hu, Liu Hui-qin, Fan Hui, Chang-Jin Kim, Y.Kitamura, Yao Kui-Ling, Fudi Li, Zhongwei Li, S.T.Ohki, S.Tajima, Shujin Wang, Dou Xintian, Ick-Dong Yoo, C.C.Young.
Europe: N.Amarger, J.E.Beringer, D.A.Bond, W.J.Broughton, R.O.Echevarrieta, K.C.Engvild, J.-C.Cleyet Marel, J.E.Cooper, C.Hera, R.Hardwick, J.Jebb, C.Knott, H.Mareckova, E.N.Mishustin, P.M.Murphy, L.R.Mytton, M.P.Nuti, P.S.Nutman, M.Obaton, A.Popescu, G.S.Posypanov, S.Redzepovic, V.Skrdleta, J.Sprent, R.J.Summerfield, R.S.Tayler, H.Tunney, J.Wery, P.M.Williams.
North Africa and West Asia: S.M.Abdel-Wahab, E.Acikgoz, R.K.Al-Rashidi, D.Beck, M.N.Ben Ali, D.Beyene, D.M.Daoud, M.Engin, M.H.Tageldin, A.Hilali, F.M.Khalifa, B.D.Kishinevsky, R.Lobel, L.A.Materon, I.Mohammed, T.M.M.Moharram, N.O.Mukhtar, M.P.Nuti, Y.Okon, I.Papastylianou, H.Saint Macary, M.S.A.Safwat, S.Sarig, M.Zaroug.
North America: M.Alexander, D.D.Baltensperger, D.F.Bezdicek, L.Bordeleau, P.J.Bottomley, B.B.Bohlool, K.W.Clark, W.R.Ellis, L.R.Frederick, D.J.Hume, W.C.Lindemann, T.Loynachan, D.N.Munns, D.A.Phillips, R.J.Rennie, P.Somasegaran, D.F.Weber, P.M.Williams, D.O.Wilson.
South America: B.Ahmed, J.I.Baldani, A.F.De Bonis, J.G.Carrion, A.A.Franco, J.M.J.Freire, C.A.Labandera Gonzalez, J.Halliday, T.A.Lie, M.Hungria, M.S.de Mallorca, S.M.T.Satio, R.Sylvester-Bradley, A.Trotman, R.Valenzuela, R.W.Weaver, D.F.Weber, P.M.Williams.
Halliday, S.V.Hegde, J.V.D.K.Kumar Rao, A.K.M.Hossain, C.M.Mudannayake, Noor Mohammad, P.T.C.Nambiar, S.Nasreen, V.Ranga Rao, H.Saint Macary, M.A.Sheikh, P.Tauro, K.D.Yami.
Southeast Asia: Faizah Abdul-Wahab, N.Boonkerd, T.P.Duong, J.Halliday, A.Ikram, Chong Kewi, T.A.Lie, I.J.Manguiat, W.M.W.Othman, B.Rerkasem, P.Singleton, Y.Taryo-Adiwiganda, Y.Vasuvat.

References

Ahmad,M.H., Eaglesham,A.R.J., Hassouna,S., Seaman,B., Ayanaba,A., Mulongoy,K. and Pulver,E.L. (1981) Examining the potential for inoculant use with cowpeas in West African soils. *Tropical Agriculture* (Trinidad), **58**, 325–335.

Ayanaba,A. (1977) Toward better use of inoculants in the humid tropics. In Ayanaba,A. and Dart,P.J. (eds), *Biological Nitrogen Fixation in Farming Systems of the Humid Tropics*. John Wiley and Sons, Chichester, pp. 181–187.

Beck,D.P. and Materon,L.A. (eds) (1988) *Nitrogen Fixation by Legumes in Mediterranean Agriculture*. Martinus Nijhoff, Lancaster.

Beijerinck,M.W. (1888) Die bacterien der Papilionaceeknollchen. *Botanische Zeitung,* **46**, 726–735, 741–750, 757–771, 781–790, 797–804.

Burton,J.C. (1982) Modern concepts in legume inoculation. In Harris,S.C. and Graham,P.H. (eds), *Biological Nitrogen Fixation Technology for Tropical Agriculture*. Centro Internacional de Agricultura Tropical (CIAT), Cali, pp. 105–114.

Chowdhury,M.S. (1977) Response of soybean to *Rhizobium* inoculation at Morogoro, Tanzania. In Ayanaba,A. and Dart,P.J. (eds), *Biological Nitrogen Fixation in Farming Systems of the Humid Tropics*. John Wiley and Sons, Chichester, pp. 245–253.

Cocks,P.S. (1988) The role of pasture and forage legumes in livestock based farming systems. In Beck,D.P. and Materon,L.A. (eds), *Nitrogen Fixation by Legumes in Mediterranean Agriculture*. Martinus Nijhoff, Lancaster, pp.3–10.

Dowling,D.N. and Broughton,W.J. (1986) Competition for nodulation of legumes. *Annual Review of Microbiology*, **40**, 131–157.
Eaglesham,A.R.J. (1985) Comparison of nodulation promiscuity of US- and Asian-type soya beans. *Tropical Agriculture* (Trinidad), **62**, 105–109.
Eaglesham,A.R.J. and Ayanaba,A. (1984) Tropical stress ecology of rhizobia, root nodulation and legume fixation. In Subba Rao,N.S. (ed.), *Current Developments in Biological Nitrogen Fixation*. Oxford and IBH Publishing Co., New Delhi, pp. 1–35.
Eaglesham,A.R.J., Hassouna,S. and Seegers,R. (1983) Effect of fertilizer N on N_2-fixation by cowpea and soybean. *Agronomy Journal*, **75**, 61–66.
Eaglesham,A.R.J., Stowers,M.D., Maina,M.L., Goldman,B.J., Sinclair,M.J. and Ayanaba,A. (1987) Physiological and biochemical aspects of diversity of *Bradyrhizobium* sp. (*Vigna*) from three west African soils. *Soil Biology and Biochemistry*, **19**, 575–581.
Eisbrenner,G. and Evans,H.J. (1983) Aspects of hydrogen metabolism in nitrogen-fixing legumes and other plant–microbe associations. *Annual Review of Plant Physiology*, **34**, 105–136.
Empresa Brasileira de Pesquisa Agropecuaria (1984) *Importance of Leguminous Trees. Pesquisa Agropecuaria Brasileira* (Special Edition) **19**, EMBRAPA.
Fred,E.B., Baldwin,I.L. and McCoy,E. (1932) *Root Nodule Bacteria and Leguminous Plants*. University of Wisconsin, Madison.
Freire,J.M.J. (ed.) (1987) *Proceedings of the Workshop on Rhizobium/Legume Inoculants*. Universidado Federal de Rio Grande do Sul, Porto Alegre.
Hardy,R.W.F. and Havelka,U.D. (1976) Photosynthate as a major factor limiting nitrogen fixation by field-grown legumes with emphasis on soybeans. In Nutman,P.S. (ed.), *Symbiotic Nitrogen Fixation in Plants*. Cambridge University Press, Cambridge, pp. 421–439.
International Institute for Tropical Agriculture (1987) *IITA Annual Report and Research Highlights 1986*. IITA, Ibadan, pp. 27–30.
Kremer,R.J. and Peterson,H.J. (1982) Effects of carrier and temperature on survival of *Rhizobium* spp. in legume inocula: development of an improved type of inoculant. *Applied and Environmental Microbiology*, **45**, 1790–1794.
Kremer,R.J. and Peterson,H.J. (1983) Field evaluation of selected *Rhizobium* in an improved legume inoculant. *Agronomy Journal*, **75**, 139–143.
Kumar Rao,J.V.D.K., Dart,P.J. and Usha Kiran,M. (1984) *Rhizobium*-induced leaf role in pigeonpea. *Soil Biology and Biochemistry*, **16**, 89–91.
La Favre,J.S. and Eaglesham,A.R.J. (1986) Rhizobitoxine: a phytotoxin of unknown function which is commonly produced by bradyrhizobia. *Plant and Soil*, **92**, 443–452.
Lambert,G.R., Harker,A.R., Zuber,M., Dalton,D.A., Hanus,F.J., Russell,S.A. and Evans,H.J. (1985) Characterization, significance and transfer of hydrogen uptake genes from *Rhizobium japonicum*. In Evans,H.J., Bottomley,P.J. and Newton,W.E. (eds), *Nitrogen Fixation Research Progress*. Martinus Nijhoff, Lancaster, pp. 209–215.
Lawes,J.B. and Gilbert,J.H. (1889) On the present position of the question of the sources of the nitrogen of vegetation, with some new results, and preliminary notice of new lines of investigation. *Philosophical Transactions of the Royal Society of London, B, Series*, **180**, 1–107.
Macary,H.S., Beunard,P., Montange,D., Tranchant,J.P. and Verniau,S. (1986) Setting and diffusion of a production system for legume *Rhizobium* inoculants. *Symbiosis*, **2**, 363–366.
Madsen,E.L. and Alexander,M. (1982) Transport of *Rhizobium* and *Pseudomonas* through soil. *Soil Science Society of America Journal*, **46**, 557–560.
Mengel,D.B., Segars,W. and Rehm,G.W. (1987) Soil fertility and liming. In Wilcox,J.R. (ed.), *Soybeans; Improvement Production and Uses*, Second edn. American Society of Agronomy, Madison, pp. 461–496.
Nangju,D. (1980) Soybean response to indigenous rhizobia as influenced by cultivar origin. *Agronomy Journal*, **72**, 403–406.
Salema,M.P., Parker,C.A., Kidby,D.K. and Chatel,D.L. (1982a) Death of rhizobia on inoculated seed. *Soil Biology and Biochemistry*, **14**, 13–14.
Salema,M.P., Parker,C.A., Kidby,D.K. and Chatel,D.L. (1982b) Study of *Rhizobium* in the legume rhizosphere. In Harris,S.C. and Graham,P.H. (eds), *Biological Nitrogen Fixation Technology for Tropical Agriculture*. Centro Internacional de Agricultura Tropical (CIAT), Cali, pp. 213–217.
Skipper,H.D., Palmer,J.H., Giddens,J.E., and Woodruff,J.M. (1980) Evaluation of commercial soybean inoculants in South Carolina and Georgia. *Agronomy Journal*, **72**, 673–674.
Smith,R.S. (1987) Production and quality of inoculants. In Elkan,G.H. (ed.), *Symbiotic Nitrogen Fixation Technology*. Marcel Dekker, New York, pp. 391–411.

Somasegaran,P. (1985) Inoculant production with diluted liquid cultures of *Rhizobium* spp. and autoclaved peat: evaluation of diluents, *Rhizobium* spp., peats, sterility requirements, storage, and plant effectiveness. *Applied and Environmental Microbiology,* **50**, 398–405.

Somasegaran,P. and Halliday,J. (1982) Dilution of liquid *Rhizobium* cultures to increase production capacity of inoculant plants. *Applied and Environmental Microbiology,* **44**, 330–333.

Somasegaran,P. and Hoben,H.J. (1985) *Methods in Legume–Rhizobium Technology,* NifTAL, Hawaii, pp. 187–200.

Summerfield,R.J. and Lawn,R.J. (1987) Tropical grain legume crops: a commentary. *Outlook on Agriculture,* **16**, 189–197.

Sylvester-Bradley,R. (1984) Rhizobium inoculation trials designed to support a tropical forage selection programme. *Plant and Soil,* **82**, 377–386.

Sylvester-Bradley,R., Mosquera,D. and Mendez,J.E. (1988) Selection of rhizobia for inoculation of forage legumes in savanna and rain forest soils of tropical America. In Beck,D.P. and Materon,L.A. (eds), *Nitrogen Fixation by Legumes in Mediterranean Agriculture.* Martinus Nijhoff, Lancaster, pp. 225–233.

Ssali,H. and Keya,S.O. (eds) (1985) *Biological Nitrogen Fixation in Africa: Proceedings of the First Conference of the African Association for Biological Nitrogen Fixation.* University of Nairobi, Nairobi.

Weaver,R.W. and Frederick,L.R. (1974) Effect of inoculum rate on competitive nodulation of *Glycine max.* L.Merrill. II. Field studies. *Agronomy Journal,* **66**, 233–236.

Williams.P.M. (1984) Current use of inoculant technology. In Alexander,M. (ed.), *Biological Nitrogen Fixation Ecology, Technology and Physiology.* Plenum Press, London, pp. 173–200.

Wynne,J.C., Bliss,F.A. and Rosas,J.C. (1987) Principles and practice of field designs to evaluate symbiotic nitrogen fixation. In Elkan,G.H. (ed.), *Symbiotic Nitrogen Fixation Technology.* Marcel Dekker, New York, pp. 371–389.

CHAPTER 4

Present and future value of mycorrhizal inoculants

D.P.STRIBLEY

AFRC Institute of Arable Crop Research, Rothamsted Experimental Station, Harpenden, Herts AL5 2JQ, UK

Introduction

In its simplest and most widely comprehended sense, the term 'mycorrhiza' implies a symbiosis between certain soil fungi and the roots of vascular plants, where the relationship is not overtly pathogenic. Mycorrhizal associations are often but not necessarily mutualistic. It is impossible to define 'mycorrhiza' more precisely, for it has become a term which now encompasses both autotrophic and saprotrophic hosts, and fungal infections of organs as diverse as orchid protocorms, rhizoids of mosses and thalli of liverworts.

Most species of vascular plants growing in the field, in all but the most extreme habitats, are mycorrhizal. It could be argued therefore that mycorrhiza is the most prevalent mutualistic symbiosis on earth; it is certainly the most widely studied. Since 1978 there have been at least 4000 publications on mycorrhiza. It may be conservatively estimated that £40m has been spent on mycorrhizal research in the last decade. The layman is entitled to ask the purpose of this not inconsiderable expenditure. Most papers have reported fundamental studies on the physiology of mycorrhizas, and on their effects on the physiology of their hosts. Since mycorrhizas, not non-symbiotic roots, are the principal organs of nutrient absorption of vascular plants, there seems ample justification for such fundamental studies. However, the discovery that, in many circumstances, mycorrhizal infection can greatly increase the rate of uptake of nutrients (particularly phosphorus and nitrogen) from deficient soils has promoted widespread debate, speculation and a certain amount of experimentation on the potential practical use of artificial inoculation with mycorrhizal fungi. These studies have been described in several detailed reviews (Rhodes, 1980; Abbott and Robson, 1982; Menge, 1983; Molina and Trappe, 1984; Powell, 1984; Creighton Miller *et al.*, 1986; Gianinazzi-Pearson, 1986; Gianinazzi and Gianinazzi-Pearson, 1986; Hayman, 1987; Howeler *et al.*, 1987; Jeffries, 1987; LeTacon *et al.*, 1987; Nemec, 1987).

This chapter does not attempt to add to this list. Rather, by concentrating on general principles, it will explore why, despite widespread interest (culminating in a major international symposium: see Sylvia *et al.*, 1988), the use of mycorrhizal inoculum has failed to make any significant impact on crop husbandry. Why is it that the promise of benefit from artificial inoculation has been largely unfulfilled? Will the situation change?

The major types of mycorrhizas will be covered, but most emphasis will be given to vesicular−arbuscular (VA) mycorrhizas, because of their wide occurrence in crop plants of major economic importance.

Biology of mycorrhiza

Some basic facts about the biology of the four types of mycorrhiza are summarized in *Table 1*. There are several excellent texts available for the reader seeking more detailed information, for example: Harley and Smith (1983); Powell and Bagyaraj (1984); Read (1984); Read and Bajwa (1985); Smith and Gianinazzi-Pearson (1988).

In VA mycorrhizas the fungal partner is restricted to the cells of the cortex where it grows both intra- and intercellularly, invaginating the host cells at intervals to form complex, dichotomously-branched structures termed arbuscules. These are thought to be the site of nutrient interchange between the symbiotic partners, but this is not yet proven. The internal mycelium is continuous with an external mycelium, which bears chlamydospores and can form interconnections with other roots. The distribution of the external mycelium, and how this is affected by the chemical and physical properties of the soil environment, is incompletely known. Infection is not systemic: it occurs *de novo* in the developing seedling from other mycorrhizas or from soil-borne propagules such as spores. There is little evidence that the fungal partner has an independent saprophytic existence in soil, and it appears to be ecologically obligate. It can be grown axenically under laboratory conditions for short periods only. The most striking property of the fungal symbiont is its apparent lack of specificity. A single isolate of a VA mycorrhizal fungus can infect a very wide range of host plants, including most important crop species, with the notable exception of most members of the Cruciferae and Chenopodiaceae (Trappe, 1987; Newman and Reddell, 1987). Vesicular−arbuscular mycorrhizas are thus economically the most important type of mycorrhiza.

The fungal partner in ectomycorrhizas forms a tight sheath round the root, and from this sheath hyphae grow into the outer cortex to form a network (the Hartig net). Hyphae connected to the sheath ramify extensively in the surrounding soil, where they may aggregate to form cords. The degree of development of this external mycelium depends on environment and fungal species. Although there are some examples of specificity between host and fungal partner in some species of ectomycorrhizal fungi, most fungi are non-specific and several different species may occur together on the same root system (Mosse *et al.*, 1981). Most of the basidiomycete and ascomycete fungi involved can be readily grown on laboratory media. The most important ectomycorrhizal hosts are the gymnospermous and angiospermous trees of temperate forests.

The roots of many plants of the family Ericaceae have a cortex which is only one or two cells thick. These so-called 'hair roots' are invaded by mycorrhizal fungi which form tight coils (pelotons) within the lumen of the cells, which are eventually killed by this invasion. There is little cell-to-cell spread of the fungus. An extensive external mycelium is formed. The endophytes of ericoid mycorrhizas are easily culturable and do not appear to be host-specific.

The symbioses formed by orchid hosts are confusing for the uninitiated for here the

Table 1. Types of mycorrhiza[a] of potential practical use.

Type	Typical host plants	Fungal partner	Culturability of fungal partner
Ecto- (or sheathing)	Forest trees (particularly Pinaceae, Fagaceae)	Basidiomycetes, Ascomycetes	Most are easily culturable on laboratory media
Vesicular–arbuscular (VA)	Most crop and ornamental plants, except for most spp. of Cruciferae, Chenopodiaceae (e.g. beets, rape)	Certain genera of the Endogonaceae (Zygomycetes)	Can only be grown for practical purposes in symbiotic association
Ericoid	Members of Ericaceae e.g. *Vaccinium*	Ascomycetes	Easily grown on wide range of standard fungal media
Orchidaceous	Orchidaceae (mainly protocorms)	Basidiomycetes	Easily grown on wide range of standard fungal media

[a] For a detailed account of the biology and classification of mycorrhizas, see Harley and Smith (1983).

term 'mycorrhiza' is applied both to fungal infection of the protocorm that develops from the germinating seed, and also to the infections of roots of adult orchids. The former association has been most intensively studied and is the most physiologically important. Young protocorms are invaded by various species of *Rhizoctonia* (imperfect stages of certain basidiomycetes) which form pelotons in the cells. Only with specific combinations of host and fungal genotype does the protocorm develop further. If the combination is wrong, the invading fungus may kill its host. The fungi are easily grown on artificial media.

Physiological effects on the host—laboratory evidence

Mycorrhizal fungi occur naturally in most soils, hence most laboratory studies have been made with soils that have been partly sterilized so that non-symbiotic control plants may be raised. It cannot be emphasized too strongly that most of our knowledge of effects of mycorrhiza upon the physiology of their hosts has been derived from pot experiments on such sterilized media. There are profound differences in chemistry and biology between sterilized soils and field soils, yet the dangers of extrapolation from pot experiments have not always been adequately appreciated (Stribley, 1987).

VA mycorrhizas

Sanders and Tinker (1973) noted that the phosphorus (P) inflow (flux per unit length of root) in non-symbiotic roots of onion was close to the maximum that could be supported by the diffusive properties of the soil. By contrast, inflow to VA mycorrhizal roots was at least 4 times greater. These authors concluded that the external hyphae of the mycorrhizas directly enhanced phosphorus uptake by absorbing it from the soil and translocating it to the host. In effect, mycorrhizal roots explored a greater volume of soil than did uninfected roots. Since this seminal paper, it has been repeatedly confirmed that VA mycorrhizas can take up phosphorus more rapidly from deficient soils than can non-symbiotic roots. Experiments with ^{32}P (Bolan *et al.*, 1984), and with soil heated to fix phosphate (Barrow *et al.*, 1977), have suggested that VA mycorrhizas do not have a disproportionate ability to take up fixed forms of phosphorus, and that mycorrhizal hyphae, like the host roots, take up phosphorus from the soil solution only. However, when Bolan *et al.* (1987) added iron oxide to soil to fix phosphorus, there was a threshold of addition below which non-mycorrhizal plants took up no phosphorus, but at which mycorrhizal plants achieved almost maximum yields. This implied that iron phosphates could be solubilized by mycorrhizas. It is important to resolve this matter, because an ability of plants to take up fixed phosphorus would greatly increase the efficiency of phosphate fertilizers, and could be an important reason to exploit VA mycorrhizal fungi. Mycorrhizal plants can grow well where rock phosphate is used as phosphorus fertilizer, but there is no evidence that mycorrhizas dissolve rock phosphate (Pairunan *et al.*, 1980). Rather, the external hyphae effectively reduce the mean distance between the particles of fertilizer and the root.

It is remarkable to note that, in the 15 years since Sanders and Tinker first propounded their simple hypothesis, there has been little conceptual advance in the understanding of the phosphorus physiology of VA mycorrhizas. We are still a long way from being able to predict how a given plant on a given soil will respond to mycorrhizal colonization. Detailed models of phosphorus uptake by non-symbiotic roots exist (Helyar and Munns, 1975; Nye and Tinker, 1977; Barber, 1984; Nye and Kirk, 1987), and it should be possible to combine these with current mathematical models of mycorrhizal colonization (Smith and Gianinazzi-Pearson, 1988) to yield a model which could predict the effect of artificial inoculation.

Mycorrhizas are not necessarily beneficial to their host. The fungal endophyte utilizes host photosynthate for growth and respiration, so it could be said that there is a 'cost' to the host for the advantage it gains in improved phosphorus uptake (Stribley *et al.*, 1980; Cooper, 1984). This 'cost' is not an entirely academic concept, for mycorrhizal plants take up more phosphorus per unit of dry matter produced than do non-mycorrhizal plants, over a wide range of concentrations of soil phosphorus (Stribley *et al.*, 1980). At high concentrations of soil phosphorus, where the host does not respond to fertilizer, there may be luxury uptake of phosphorus in mycorrhizal plants (Smith *et al.*, 1986), and reductions in yield, presumably attributable to the carbon drain (Cooper, 1984).

Plants grown on phosphorus-deficient soil and artificially inoculated with VA mycorrhizal fungi show many physiological changes. It is important to know whether such changes are simply indirect effects of improved phosphorus nutrition, or are unique consequences of mycorrhizal infection. Particularly important with regard to potential practical use are changes in uptake of metals, in water relations and disease resistance. It is now well established that in pot experiments VA mycorrhizas can directly enhance the uptake of zinc, copper and possibly iron (Gildon and Tinker, 1983b; Tinker and Gildon, 1983; Pacovsky, 1986; Kucey and Janzen, 1987). This seems to arise because these elements, like phosphorus, are slow to diffuse through soil and form a depletion zone round roots which can be short-circuited by the external hyphae (Tinker and Gildon, 1983). Paradoxically, there is evidence that VA mycorrhizas can inhibit uptake of zinc, at toxic concentrations in soil, thus reducing adverse effects on the host (Dueck *et al.*, 1986). Evidence is also accumulating that VA mycorrhizas can protect their hosts from toxic concentrations of manganese (Pacovsky, 1986). This finding is serious because soluble manganese is released from soil by most sterilization procedures, and one may speculate that many pot experiments on effects of mycorrhizas on phosphorus uptake have been confounded by these manganese effects. Killham and Firestone (1983) found that mycorrhizal colonization increased uptake by the grass *Ehrharta calycina* of heavy metals applied in simulated rain, at rates calculated to be similar to those resulting from smelter effluents. These contradictory studies suggest that the involvement of VA mycorrhizas in metal uptake requires further study.

In a careful review of effects of VA mycorrhizas on water relations of their hosts, Nelsen (1987) concluded that most effects are probably an indirect consequence of improved phosphorus nutrition. The experiment of Fitter (1988), in which a factorial combination of phosphorus treatments and inoculation treatments was used, strongly supports this view. Fitter comments on the disturbing inconsistency of experimental results on water relations of VA mycorrhizas, and shows that most investigations on this topic have not completely taken into account the confounding effects of phosphorus nutrition.

Effects of colonization by VA mycorrhizal fungi on resistance to disease caused by root-invading pathogens (particularly fungi and nematodes) are inconsistent. They depend on complex circumstances such as the timing of addition of inoculum of the pathogens in relation to that of the mycorrhizal fungi. There is little clear evidence, except perhaps from the studies of Caron et al. (1985) and of Kaye et al. (1894) on the fungal pathogens *Fusarium oxysporum* and *Pythium ultimum*, respectively, of a direct effect of VA mycorrhizal fungi in preventing ingress of pathogens. Once again, most of these effects of VA mycorrhizal fungi can be probably explained in terms of an indirect effect of phosphorus nutrition (Smith, 1988).

Ectomycorrhizas

The major physiological effect of ectomycorrhizas, as in VA mycorrhizas, seems to be to improve uptake of phosphorus (Harley and Smith, 1983; Finlay and Read, 1986). There is also growing evidence that external mycelium of ectomycorrhizas can mineralize organic forms of nitrogen, such as proteins, and thereby improve access of the host to nitrogen in organic soils (Abuzinadah et al., 1986). This work supports an hypothesis first propounded almost a century ago (Frank, 1894)!

There is good evidence that ectomycorrhizas can protect their hosts from adverse effects of heavy metals in soil. Dixon and Buschena (1988) found that ectomycorrhizal *Pinus* and *Picea* were protected against low or intermediate concentrations of cadmium, copper, nickel, lead and zinc. An environment which contains heavy metals may select out metal-tolerant strains of ectomycorrhizal fungi, which in turn may protect the host against uptake of those heavy metals (Colpaert and Van Assche, 1987).

Ectomycorrhizas alter the water relations of their hosts. For example, seedlings of Douglas fir inoculated with *Rhizopogon vinicola* were more resistant to drought than were non-mycorrhizal seedlings (Parke et al., 1983). But again it is probable that most effects on water relations are a secondary consequence of improved mineral nutrition, and the burden of proof of a unique mycorrhizal effect must rest with its proponents.

The evidence that ectomycorrhizas can protect the host against disease by root-invading pathogens is much stronger than it is for VA mycorrhizas. For example, Chakravarty and Unestam (1987) found that species of ectomycorrhizal fungi would confer resistance in *Pinus sylvestris* seedlings to *Fusarium moniliforme* and *Rhizoctonia*, but the degree of resistance conferred, and its longevity, depended strongly upon the genotype of the mycorrhizal fungus.

Ericoid mycorrhizas

Extensive work by Read and co-workers has demonstrated that ericoid mycorrhizas are important in the nitrogen nutrition of their hosts (Read and Bajwa, 1985). The external mycelium has proteolytic capability and may render available to the host plant nitrogen from the organic soils on which ericaceous species normally grow. Mycorrhizal colonization can, as in VA mycorrhizas and ectomycorrhizas, reduce uptake of toxic metals by the host (Bradley et al., 1981).

Orchid mycorrhizas

Under natural conditions, orchid protocorms will not develop unless infected by symbiotic fungi. The mechanism for this stimulation of growth is unclear but may involve transfer of carbon compounds from the external substrate. Under artificial conditions, the activity of the symbiotic fungus can be replaced by suitable media containing a mixture of sugars and salts (Hadley, 1982).

Mycorrhizal colonization of roots of adult orchids may be important for adequate phosphorus nutrition of the host, but much more work is needed to elucidate this (Harley and Smith, 1983).

Why use inoculants?

The rate of colonization of roots by mycorrhizal fungi is an important determinant of host response, particularly in annual plants, and this in turn is strongly controlled by the concentration of inoculum in the substrate. Although mycorrhizal fungi in general are widespread in soils, artificial inoculation could improve host response where the concentration of natural inoculum is suboptimal. In special instances, such as in fumigated nursery beds, or where soil-less media are used in the glasshouse industry, mycorrhizal inoculum may be totally absent. There is also the possibility of supplementing or supplanting natural fungi with isolates that are innately superior in their physiological effects on the host.

VA mycorrhiza

Natural inoculum of VA mycorrhizal fungi is present in most agricultural soils but there is an embarrassing lack of information about its properties and distribution, and how these are affected by agricultural practices. Sanders and Sheikh (1983) reported that concentrations of propagules of VA mycorrhizal fungi in arable topsoils in the Leeds area, UK, estimated by the most probable number technique, lay in the range $0.1-10$ propagules g^{-1} of soil. This concentration may be suboptimal, for in a pot experiment with maize, Sanders and Sheikh (1983) showed that 25 spores g^{-1} of soil were necessary to give maximum rate of development of infected roots. Tillage may be an important factor governing the concentration and infectivity of inoculum. Evans and Miller (1988) and Fairchild and Miller (1988) showed that disturbance of soil cores taken from a zero till system markedly reduced the rate of development of VA mycorrhizas in maize sown in them, and this in turn reduced absorption of phosphorus. Tillage has been reported to reduce infection in field-grown maize (Anderson *et al.*, 1987). Mechanical damage to mycorrhizal roots and hyphae presumably accounted for the reduction in infectivity but the mechanisms are not fully understood. In the field this effect may be exacerbated by inversion of the topsoil, where the highest concentration of mycorrhizas occurs, by ploughing.

The concentration of inoculum in soil can be roughly inferred from the time-course of colonization of roots of field crops by indigenous fungi. The literature on this, however, is remarkably inconsistent. In a major survey of the effect of agronomic factors and soil conditions on mycorrhizal development in barley (*Hordeum vulgare*), Black and Tinker (1979) found a long delay before an appreciable fraction of the root system became infected. By contrast, Dodd and Jeffries (1986) showed that in three sites in South East England over three successive seasons, there was a peak of infection (of up to 60% root length infected) in November to December. Their technique differed from that of Black and Tinker (1979) in that a compound microscope (at ×100 magnification), rather than a low-power dissecting microscope, was used to observe infection in roots.

Differences in efficacy of strains have rarely been demonstrated, because in most studies there have been confounding effects of inoculum concentration. Haas and Krikun (1985) clearly demonstrated intraspecific differences in efficacy but even here these effects were minor compared to those of concentration of inoculum. Almost nothing is known about genetic variation in VA mycorrhizal fungi, or about the effects of environment upon natural selection of genotypes. However, it is clear that certain genotypes are better adapted than others to major variables in soil chemistry such as pH (Wang *et al.*, 1985), and there is considerable scope for empirical selection of genotypes for particular environments. Artificial inoculation may be unnecessary if an effective population of native mycorrhizal fungi can be built up by management practices such as precropping with suitable host species. Such possibilities have been demonstrated in field trials in the eastern plains of Colombia (I.Arias, J.M.Day, J.C.Dodd, D.S.Hayman and I.Koomen, pers. comm.). These possibilities should be urgently investigated, particularly in respect of reclamation of soils for tropical agriculture.

Ectomycorrhizas

Bare root seedlings which are used in temperate forestry are normally raised in nurseries where there may be cultural practices (such as fumigation and the use of biocides) inimical to the development of ectomycorrhizas (Molina and Trappe, 1984). When nurseries are established on sites remote from native forests, the soil may also lack suitable inoculum.

In tropical forestry, foreign species of tree introduced for the purposes of afforestation may fail to establish because of the lack of suitable inoculum. For example, attempts to introduce Monterey pine (*Pinus radiata*) into Australia and New Zealand failed until mycorrhizal fungi from the native habitat of the pine were accidentally introduced. However, in such cases inoculum may need only to be introduced once because of the extreme efficiency with which most (basidiomycetous) ectomycorrhizal fungi spread by means of spores.

Ericoid mycorrhizas

There would appear to be little need for artificial inoculation endophytes of ericoid mycorrhizas. Pearson and Read (1973) found that *Hymenoscyphus ericae* was present in a normal garden soil remote from any heathland environment. The other important endophytes, *Oidiodendron* spp. (Dalpé, 1986), are very common saprophytes found in a wide variety of habitats.

Practical trials with mycorrhizal inoculants

VA mycorrhizas

Inoculum of VA mycorrhizal fungi has been tested with a wide range of crops and cropping systems (*Table 2*). Because large-scale production of the symbiotic fungus

Table 2. Crops and cropping systems[a] used in tests of practical use of inoculum of vesicular–arbuscular mycorrhizal fungi.

System	Examples of crops tested
Field crops	
(i) Temperate	
Arable crops	Small-grain cereals (wheat, barley)
	Large-grain cereals (maize)
	Potatoes
Vegetable crops	Transplanted (onion, leek, asparagus, bell pepper)
	Drilled (peas, onion)
Upland pasture	Clover and grasses
Fruit crops	Citrus (raised on fumigated nursery soils)
Forestry	Hardwood species (*Acer, Fraxinus*)
(ii) Tropical	
Arable crops	Small grain cereals (wheat, sorghum)
	Root crops (cassava)
	Cash crops (coffee, tea)
	Legumes (soya)
Pasture	Legumes (kudzu) and grasses
Hardy nursery stock	
Ornamentals	*Pelargonium, Fucshia, Juniperus*

[a]This table is not intended to be comprehensive: for full information the reader should consult the review articles mentioned in the text.

in axenic medium is not yet possible, the inocula used in the field trials have all been derived from open pot cultures with suitable host plants, and hence there has always been the possibility that pathogens have accidentally been introduced. Commercial inocula have been produced and marketed to a limited extent in the USA (Jeffries, 1987). The difficulty of producing reliable inoculum of VA mycorrhizal fungi has undoubtedly restricted the scale and scope of field trials. Many trials have been criticized on the grounds that unrealistic quantities of inoculum were used. This criticism seems unjustified, however, because the experiments were tests of principle rather than demonstrations of immediate commercial application. Prospects for production of pathogen-free and easily handled inoculum have now improved greatly with the introduction of the use of calcined montmorillonite clays as the growth medium in the mother cultures (Dehne and Backhaus, 1986). This type of inoculum is currently being actively developed by the Agricultural Genetics Company in the UK.

It would be unrealistic and uneconomic to attempt to use artificial inoculation to increase the concentration of inoculum throughout the soil profile. In most studies, inoculum has not been broadcast but has been placed in close contact with roots of developing seedlings (Hayman et al., 1981; Powell, 1984). Since the linear rate of extension of mycorrhizal fungi along roots is slower than the rate of extension of the roots themselves (Amijee et al., 1986), the inoculant fungus becomes mainly confined to proximal parts of the root system, and the distal parts extend into soil containing natural inoculum. In laboratory experiments, Hepper et al. (1988) have shown that the naturally occurring fungus can strongly inhibit spread from the localized inoculum. Such competition may have caused failure of artificial inoculation in field experiments.

In general, results from experiments in temperate arable systems have been inconsistent and conflicting (Fitter, 1985). Work with cereals, the crop with potentially the largest application for mycorrhizal inoculants, has been particularly disappointing. Buwalda et al. (1985a,b) grew spring wheat on plots that were deficient in phosphate and which had been fallowed and recently limed. Under these conditions, there was a large potential for response to inoculant fungi. Added inoculum gave a small increase in yield of grain on fumigated plots but had little effect on non-fumigated plots even though the wheat responded well to added phosphorus. In some experiments, there were increases in concentration of phosphorus in shoots of inoculated plants that did not show increased growth, indicating that, in some circumstances, mycorrhizal inoculation may *decrease* efficiency of fertilizer use. Buwalda et al. (1985a,b) concluded that wheat and barley are inherently unresponsive to mycorrhizal colonization. In the experiments of Baltruschat (1987a,b), inoculation of maize with high-quality inoculum increased growth at the 5−6 leaf stage only when inoculum was directly incorporated into seed rows and phosphorus fertilizer was placed underneath the seeds. By harvest, however, there were no effects of inoculation on yields of cob or straw.

In trials where increases in crop yields have been demonstrated, the effects appear to have been a direct consequence of improved phosphorus nutrition. There appear to have been no unequivocal demonstrations of effects on water relations, uptake of trace metals or disease-tolerance. Snellgrove and Stribley (1986) showed that inoculation of transplanted onions improved yields because a greater proportion of plants in this treatment formed bulbs. This effect appeared to be unrelated to phosphorus nutrition but needs verification. Schüepp et al. (1987) demonstrated that lettuce and maize took up less heavy metals from contaminated soil when inoculated with VA mycorrhizal fungi.

Experiments in the tropics have been more encouraging. Inoculation increased growth of cassava (*Manihot esculenta*) and legumes such as kudzu (*Pueraria phaseoloides*) in soils low in natural inoculum (Howeler et al., 1987). Other trials have shown that a combination of the use of mycorrhizal inoculants and rock phosphate could give an effective and cheap way of reclaiming the acid soils of savannahs for food production (Hammond et al., 1986; Arias et al., pers. comm.).

Ectomycorrhizas

Considerable success has been achieved in the large-scale production of inoculum of ectomycorrhizal fungi (LeTacon et al., 1987). The inoculum may be grown in fermenters and incorporated on solid substrates such as peat moss/vermiculite. Companies in the USA and France are currently developing ectomycorrhizal inocula.

There have been many demonstrations in nurseries in the USA, particularly by Marx and co-workers, mainly using the fungus *Pisolithus tinctorius*, that inoculation of nursery beds can improve the vigour and success of transplanted seedlings (Jefferies, 1987). However, it is not clear to what extent fertilizers could be used to achieve the same end, and what the relative cost would be. Furthermore, growers often seem to achieve acceptable results by using crude inoculum from forest soils. So, despite the very large literature and research expenditure on ectomycorrhizal inoculum, little impact seems to have been made on normal forestry practice.

Ericoid mycorrhizas

Powell and Bates (1981) demonstrated that blueberries (*Vaccinium corymbosum*), grown on land previously maintained as pasture, responded to inoculation with the ericoid mycorrhizal fungus *Hymenoscyphus ericae*. Increases in fruit yield of 4-year-old plants ranged from 11 to 92% depending on the cultivar. A commercial formulation of a *Hymenoscyphus* inoculum is marketed as Mycoaid by Mintech Ltd of New Zealand.

Orchid mycorrhizas

Most commercially valuable orchids can easily be propagated asymbiotically, and hence at present there appears to be little scope for use of mycorrhizal fungi (Hadley, 1982). The symbiotic method of raising orchids from seed may, however, be important in conservation projects. This has been demonstrated at the Royal Botanic Gardens, Kew, where use of mycorrhizal fungi enabled the propagation from seed of British and European orchids which are threatened in the wild (Clements *et al.*, 1986).

Conclusions

There is little evidence that the major types of mycorrhizas (VA and ecto-) have any significant effects on the physiology of their hosts other than those associated with improved uptake of phosphorus and nitrogen. Putatively unique effects of mycorrhizas have been difficult to demonstrate consistently, even under controlled conditions, and they remain controversial. The essential question regarding the potential use of mycorrhizal inoculants is, therefore, are there any practical benefits in using inoculants which allow crops to be grown at lower concentrations of phosphorus (and possibly nitrogen) in the soil? The environmental problem can be disposed of fairly easily: phosphate fertilizers are strongly retained by the soil profile and do not constitute a long-term environmental hazard (Emsley and Hall, 1976). The economic implications require more thought. The economics should form an inextricable part of the aims and logic of experiments with inoculants, but unfortunately, except in a very few cases, they have not been taken into account. Much field work with mycorrhizal inoculants has not only been poorly executed (Powell, 1984), but has also been undertaken with no clear aim in mind. The consequence has been that many studies have been neither good applied science, nor good fundamental work, and most experiments have merely demonstrated under field conditions what is known from work in pots. Much greater collaboration between those working with mycorrhiza, agronomists and economists is clearly indicated.

It is difficult to see any economic advantage in using VA mycorrhizal inoculants on crops of high value. Here, the cost of the phosphate fertilizer is trivial compared to the cash value of the crop. For example, for bulb onions the *total* fertilizer (N, P, K) cost represents only 16% of the variable costs, and a mere 7% of the gross margin (Nix, 1987). Superphosphate is extremely efficient: it has been available for 140 years, it works very well and predictably, and is unlikely to be superseded. This must be set against the complexity and unreliability of exploiting a biological system; the mycorrhizal symbiosis. However, where the economic value of the crop is low and much phosphate needs to be added to the soil, as in the reclamation of the large areas of acid phosphate-deficient soils of the tropics (Goedert, 1983), then use of mycorrhizal inoculants may become a practical and economic reality. However, even here it may be that artificial inoculants are unnecessary, and that manipulation of natural inoculum by cropping practices may be adequate.

It could be argued that it is dangerously arrogant to make pronouncements on fertilizer use which are based on today's economic and social criteria. Woolhouse (1988) has opined: 'I am convinced that we shall be in a position to look back 25 years from now on the current era of crop production based heavily on chemicals, as one which was generally effective, but relatively crude, expensive and environmentally unacceptable'. Although world reserves of phosphate rock are extremely large (Emsley and Hall, 1976), they are finite, and it is conceivable that efficient use of phosphorus fertilizers may become an important consideration.

Inefficiency of phosphorus fertilizer arises, as Tinker (1985) has pointed out, largely from the need to add large amounts of phosphorus to deficient soils to obtain maximum crop yields. This is because much added phosphorus is rapidly converted into non-labile forms (Johnston and Poulton, 1977; Barrow, 1980), which are poorly soluble and cannot be fully recovered if high crop yields are to be maintained (Mattingley and Widdowson, 1963; Johnston *et al.*, 1986). Earlier and more extensive colonization of root systems with VA mycorrhizal fungi would enable maximum yields to be attained at lower equilibrium concentrations of phosphorus in the soil solution, and could thus improve recovery of non-labile phosphorus, or reduce the necessity for high initial dressings of phosphorus fertilizer in virgin soils.

At present, however, the prospects for use of artificial inoculum of VA mycorrhizal fungi seem very circumscribed. If the claimed effects on flowering of ornamental plants (Backhaus, 1983) can be verified, there may be a role in horticulture, and possibly also in organically grown vegetables, for here the phosphorus fertilizers permitted by the Organic Standards Committee are poorly soluble.

There is possibly some promise in exploiting VA mycorrhizas to control uptake of metals by crops. Pot experiments have provided strong evidence that VA mycorrhizas increase uptake of soil copper and zinc by the host. But it is uncertain whether such effects would be of practical benefit in the field. There is currently much interest in the clinical effects of zinc deficiency in foodstuffs, and there is evidence that heavy applications of phosphorus fertilizer may reduce zinc uptake by crops (Bryce-Smith and Hodgkinson, 1986). Elimination of mycorrhizas by phosphorus fertilizers may be partly responsible for this reduced uptake (Singh *et al.*, 1986), which could perhaps be corrected by artificial inoculation. However, trace element deficiencies are easily corrected by chemical treatment (Tinker, 1986) so it is probably not presently worth the time and trouble of using inoculants. Reduction in uptake of some metals by mycorrhizal plants is more interesting. Recently Moore (1988) argued that metal-tolerant mycorrhizal inoculants could be used to reduce uptake of crops of heavy metals in sewage sludge, thereby increasing the scope for recycling the nutrient in this waste material. Almost certainly, appropriate VA mycorrhizal fungi (Spitko and Manning, 1981; Gildon and Tinker, 1983a; Angle and Heckman, 1986) could be used for this purpose (Schüepp et al., 1987), but it is probably much more desirable to attack the problem by preventing industrial wastes from reaching sewage plants.

Prospects for the use of ectomycorrhizal inoculum in establishing tropical forests in reclaimed areas and for establishing trees on difficult sites such as mine spoils seem to be quite good. But it remains unproven whether mycorrhizal fungi have any benefits in temperate forestry practice that cannot be more reliably realized by use of chemical fertilizers and careful husbandry. It seems unlikely that ericoid mycorrhizal fungi will find practical use on any large scale because of the ubiquity of natural inoculum and the high nutrient levels at most commercial sites. Fungi of orchid mycorrhizas will probably not be exploited commercially, given the simplicity and success of the non-symbiotic methods for germination of protocorms. However the successes of the Kew project, in obtaining rapid growth of symbiotically infected protocorms of difficult subjects, point the way to at least some future practical exploitation of orchid mycorrhizal fungi.

The overall picture that emerges is that the prospects for practical application of mycorrhizal inoculants are very limited both at present and for the foreseeable future. Applications may increase if efficient use of phosphorus fertilizers becomes an important consideration, but this is a matter of conjecture. In the long term, it may be simpler and more reliable to improve efficiency of fertilizer use by plants through breeding (Graham, 1984). The question posed at the beginning of this chapter can now be answered: mycorrhizal inoculants have failed to fulfil their promise because currently there is little promise to fulfil.

References

Abbott,L.K. and Robson,A.D. (1982) The role of vesicular-arbuscular mycorrhizal fungi in agriculture and the selection of fungi for inoculation. *Australian Journal of Agricultural Research*, **33**, 389–408.

Abuzinadah,R.A., Finlay,R.D. and Read,D.J. (1986) The role of proteins in the nitrogen nutrition of ectomycorrhizal plants. II. Utilization of protein by mycorrhizal plants of *Pinus contorta*. *New Phytologist*, **103**, 495–506.

Amijee,F., Stribley,D.P. and Tinker,P.B. (1986) The development of endomycorrhizal root systems. VI. The relationship between development of infection, and intensity of infection in young leek roots. *New Phytologist*, **102**, 293–301.

Anderson,E.L., Millner,P.D. and Kuniski,H.M. (1987) Maize root length density and mycorrhizal infection as affected by tillage and soil phosphorus. *Journal of Plant Nutrition*, **10**, 1349–1356.

Angle,J.S. and Heckman,J.R. (1986) Effect of soil pH and sewage sludge on VA mycorrhizal infection of soybeans. *Plant and Soil*, **93**, 437–441.

Backhaus,G.F. (1983) Einfluss der vesikulären-arbusculären Mycorrhiza auf die generative Entwicklung von *Heliotropum* und *Fuschia*. *Gartenbauwissenschaft*, **48**, 197–201.

Baltruschat,H. (1987a) Evaluation of the expanded clay as a carrier material for VA mycorrhiza spores in field inoculation of maize. *Angewandte Botanik*, **61**, 163–169.

Baltruschat,H. (1987b) Field inoculation of maize with vesicular-arbuscular mycorrhizal fungi by using expanded clay as a carrier material for mycorrhiza. *Zeitschrift für Pflanzenkrankheiten und Pflanzenschutz*, **94**, 419–430.

Barber,S.A. (1984) *Soil Nutrient Bioavailability*. John Wiley, New York.
Barrow,N.J. (1980) Evaluation and utilization of residual phosphorus in soils. In Khasawneh,F.E., Sample,E.C. and Kamprath,E.J. (eds), *The Role of Phosphorus in Agriculture*. American Society of Agronomy, Madison, pp. 333–359.
Barrow,N.J., Malajczuk,N. and Shaw,T.C. (1977) A direct test of the ability of vesicular-arbuscular mycorrhiza to help plants take up fixed soil phosphate. *New Phytologist*, **78**, 269–276.
Black,R. and Tinker,P.B. (1979) The development of endomycorrhizal root systems. II. Effect of agronomic factors and soil conditions on the development of vesicular-arbuscular mycorrhizal infection in barley and on the endophyte spore density. *New Phytologist*, **83**, 401–413.
Bolan,N.S., Robson,A.D., Barrow,N.J. and Aylmore,L.A.G. (1984) Specific activity of phosphorus in relation to the availability of phosphorus to plants. *Soil Biology and Biochemistry*, **16**, 299–304.
Bolan,N.S., Robson,A.D. and Barrow,N.J. (1987) Effects of vesicular-arbuscular mycorrhiza on the availability of iron phosphates to plants. *Plant and Soil*, **99**, 401–410.
Bradley,R., Burt,A.J. and Read,D.J. (1981) The biology of mycorrhiza in the Ericaceae. VIII. The role of mycorrhizal infection in heavy metal resistance. *New Phytologist*, **91**, 197–209.
Bryce-Smith,D. and Hodgkinson,L. (1986) *The Zinc Solution*. Arrow Books, London.
Buwalda,J.G., Stribley,D.P. and Tinker,P.B. (1985a) Effects of vesicular-arbuscular mycorrhizal infection in first, second and third cereal crops. *Journal of Agricultural Science (Cambridge)*, **105**, 631–647.
Buwalda,J.G., Stribley,D.P. and Tinker,P.B. (1985b) Vesicular-arbuscular mycorrhizae of winter and spring cereals. *Journal of Agricultural Science (Cambridge)*, **105**, 649–657.
Caron,M., Fortin,J.A. and Richard,C. (1985) Influence of substrate on the interaction of *Glomus intraradices* and *Fusarium oxysporum* f.sp. *radicis-lycopersici* on tomatoes. *Plant and Soil*, **87**, 233–239.
Chakravarty,P. and Unestam,T. (1987) Differential influence of ectomycorrhizae on plant growth and disease resistance in *Pinus sylvestris* seedlings. *Journal of Phytopathology*, **120**, 104–120.
Clements,M.A., Muir,H.and Cribb,P.J. (1986) A preliminary report on the symbiotic germination of European terrestrial orchids. *Kew Bulletin*, **41**, 437–445.
Colpaert,J.V. and Van Assche,J.A. (1987) Heavy metal tolerance in some ectomycorrhizal fungi. *Functional Ecology*, **1**, 415–421.
Cooper,K.M., (1984) Physiology of VA mycorrhizal associations. In Powell,C.Ll. and Bagyaraj,D.J. (eds), *VA Mycorrhiza*. CRC Press, Boca Raton, pp. 155–186.
Creighton Miller,J., Jr., Rajapakse,S. and Garber,R.K. (1986) Vesicular-arbuscular mycorrhizae in vegetable crops. *Hortscience*, **21**, 974–984.
Dalpé,Y. (1986) Axenic synthesis of ericoid mycorrhiza in *Vaccinium angustifolium* Ait. by *Oidiodendron* species., *New Phytologist*, **103**, 391–396.
Dehne,H.W. and Backhaus,G.F. (1986) The use of vesicular-arbuscular mycorrhizal fungi in plant production. I. Inoculum production. *Zeitschrift für Pflanzenkrankheiten und Pflanzenschutz*, **93**, 415–424.
Dixon,R.K. and Buschena,C.A. (1988) Response of ectomycorrhizal *Pinus banksiana* and *Picea glauca* to heavy metals in soil. *Plant and Soil*, **105**, 265–271.
Dodd,J.C. and Jeffries,P. (1986) Early development of vesicular-arbuscular mycorrhizas in autumn-sown cereals. *Soil Biology and Biochemistry*, **18**, 149–154.
Dueck,Th.A., Bisser,P., Ernst,W.H.O. and Schat,H. (1986) Vesicular-arbuscular mycorrhizae decrease zinc-toxicity to grasses growing in zinc-polluted soil. *Soil Biology and Biochemistry*, **18**, 331–333.
Emsley,J. and Hall,D. (1976) *The Chemistry of Phosphorus*. Harper and Row, London.
Evans,D.G.and Miller,M.H. (1988) Vesicular-arbuscular mycorrhizas and the soil-disturbance-induced reduction of nutrient absorption in maize. I. Causal relations. *New Phytologist*, **110**, 67–74.
Fairchild,G.L. and Miller,M.H. (1988) Vesicular-arbuscular mycorrhizas and the soil-disturbance-induced reduction of nutrient absorption in maize. II. Development of the effect. *New Phytologist*, **110**, 75–84.
Finlay,R.D. and Read,D.J. (1986) The uptake and distribution of phosphorus in ectomycorrhizal mycelium. In Gianinazzi-Pearson,V. and Gianinazzi,S. (eds), *Physiological and Genetical Aspects of Mycorrhizae*. INRA-Presse, Paris, pp. 351–355.
Fitter,A.H. (1985) Functioning of vesicular-arbuscular mycorrhiza under field conditions. *New Phytologist*, **99**, 257–265.
Fitter,A.H. (1988) Water relations of red clover *Trifolium pratense* L. as affected by VA mycorrhizal infection and phosphorus supply before and during drought. *Journal of Experimental Botany*, **39**, 595–603.
Frank,A.B. (1894) Die Bedeutung der Mykorrhiza-Pilze für die gemeine Kiefer. *Forstwissenschaftliches Zentralblatt*, **16**, 1852–1890.
Gianninazzi-Pearson,V. (1986) Mycorrhizae: a potential for better use of phosphate fertilizer. *Fertilizers and Agriculture*, **92**, 3–12.

Gianinazzi,S. and Gianinazzi-Pearson,V. (1986) Progress and headaches in endomycorrhiza biotechnology. *Symbiosis*, **2**, 139–149.

Gildon,A. and Tinker,P.B. (1983a) Interactions of vesicular-arbuscular mycorrhizal infection and heavy metals in plants. I. The effects of heavy metals on the development of vesicular-arbuscular mycorrhizas. *New Phytologist*, **95**, 247–261.

Gildon,A. and Tinker,P.B. (1983b) Interactions of vesicular-arbuscular mycorrhizal infection and heavy metals in plants. II. The effects of infection on uptake of copper. *New Phytologist*, **95**, 263–268.

Goedert,W.J. (1983) Management of the Cerrado soils of Brazil; a review. *Journal of Soil Science*, **34**, 405–428.

Graham,R.D. (1984) Breeding for nutritional characteristics in cereals. In Tinker,P.B. and Lauchli,A. (eds), *Advances in Plant Nutrition*, Vol. 1. Praeger, New York, pp. 57–102.

Haas,J.H. and Krikun,J. (1985) Efficacy of endomycorrhizal-fungus isolates and inoculum quantities required for growth response. *New Phytologist*, **100**, 613–621.

Hadley,G. (1982) *Orchid mycorrhiza*. In *Orchid Biology. Reviews and Perspectives*, vol. II, Cornell University Press, Ithaca, pp. 83–118.

Hammond,L.L., Chien,S.H. and Mokwunye,A.U. (1986) Agronomic value of unacidulated and partially acidulated phosphate rocks indigenous to the tropics. *Advances in Agronomy*, **40**, 89–140.

Harley,J.L. and Smith,S.E. (1983) *Mycorrhizal Symbiosis*. Academic Press, London.

Hayman,D.S. (1987) VA mycorrhizas in field crop systems. In Safir,G.R. (ed.), *Ecophysiology of VA Mycorrhizal Plants*. CRC Press, Boca Raton, pp. 193–205.

Hayman,D.S., Morris,E.J. and Page,R.J. (1981) Methods for inoculating field crops with mycorrhizal fungi. *Annals of Applied Biology*, **99**, 247–253.

Helyar,K.R. and Munns,D.N. (1975) Phosphate fluxes in the soil-plant system: a computer simulation. *Hilgardia*, **43**, 103–130.

Hepper,C.M., Azcon-Aguilar,C., Rosendahl,S. and Sen,R. (1989) Competition between three species of *Glomus* used as spatially separated introduced and indigenous mycorrhizal inocula for leek (*Allium porrum* L.). *New Phytologist*, (in press).

Howeler,R.H., Sieverding,E. and Saif,S. (1987) Practical aspects of mycorrhizal technology in some tropical crops and pastures. *Plant and Soil*, **100**, 249–283.

Jeffries,P. (1987) Use of mycorrhizae in agriculture. *CRC Critical Reviews in Biotechnology*, **5**, 319–357.

Johnston,A.E., Lane,P.W., Mattingly,G.E.G., Poulton,P.R. and Hewitt,M.V. (1986) Effects of soil and fertilizer P on yields of potatoes, sugar beet, barley and winter wheat on a sandy clay loam soil at Saxmundham, Suffolk. *Journal of Agricultural Science (Cambridge)*, **106**, 155–167.

Johnston,A.E. and Poulton,P.R. (1977) Yields on the Exhaustion Land and changes in the NPK content of the soil due to cropping and manuring, 1852–1975. *Rothamsted Experimental Station, Report for 1976*, Part 2, pp. 53–85.

Kaye,J.W., Pfleger,F.L. and Stewart,E.L. (1984) Interaction of *Glomus fasciculatum* and *Pythium ultimum* on greenhouse-grown poinsettia. *Canadian Journal of Botany*, **62**, 1575–1579.

Killham,K. and Firestone,M.K. (1983) Vesicular-arbuscular mycorrhizal medication of grass response to acidic and heavy metal depositions. *Plant and Soil*, **72**, 39–48.

Kucey,R.M.N. and Janzen,H.H. (1987) Effects of VAM and reduced nutrient availability on growth and phosphorus and micronutrient uptake of wheat and field beans under greenhouse conditions. *Plant and Soil*, **104**, 71–78.

LeTacon,F., Garbaye,J. and Carr,G. (1987) The use of mycorrhizas in temperate and tropical forests. *Symbiosis*, **3**, 179–206.

Mattingly,G.E.G. and Widdowson,F.V. (1963) Residual value of superphosphate and rock phosphate on an acid soil. I. Yields and phosphorus uptakes in the field. *Journal of Agricultural Science, Cambridge*, **60**, 399–407.

Menge,J.A. (1983) Utilization of vesicular-arbuscular mycorrhizal fungi in agriculture. *Canadian Journal of Botany*, **61**, 1015–1024.

Molina,R. and Trappe,J.M. (1984) Mycorrhiza management in bareroot nurseries. In Duryea,M.L. and Landis,T.D. (eds), *Forest Nursery Manual*. Martinus Nijhoff, The Hague, pp. 211–223.

Moore,P.D. (1988) Essential elements from waste. *Nature*, **333**, 706.

Mosse,B., Stribley,D.P. and LeTacon,F. (1981) Ecology of mycorrhizae and mycorrhizal fungi. *Advances in Microbial Ecology*, **5**, 137–210.

Nelsen,C.E. (1987) The water relations of vesicular-arbuscular mycorrhizal systems. In Safir,G.R. (ed.), *Ecophysiology of VA Mycorrhizal Plants*. CRC Press, Boca Raton, pp. 71–91.

Nemec,S. (1987) VA mycorrhizae in horticultural systems. In Safir,G.R. (ed.), *Ecophysiology of VA Mycorrhizal Plants*. CRC Press, Boca Raton, pp. 193–205.

Newman,E.I. and Reddell,P. (1987) The distribution of mycorrhizas among families of vascular plants. *New Phytologist*, **106**, 745–751.

Nix,J. (1987) *Farm Management Pocketbook* (17th edn). Wye College, University of London.

Nye,P.H. and Kirk,G.J.D. (1987) The mechanism of rock phosphate solubilization in the rhizosphere. *Plant and Soil*, **100**, 127–134.

Nye,P.H. and Tinker,P.B. (1977) *Solute Movement in the Soil–Root System*. Blackwell Scientific, Oxford.

Pacovsky,R.S. (1986) Micronutrient uptake and distribution in mycorrhizal or phosphorus-fertilized soybeans. *Plant and Soil*, **95**, 379–388.

Pairunan,A.K., Robson,A.D. and Abbott,L.K. (1980) The effectiveness of vesicular-arbuscular mycorrhizas in increasing growth and phosphorus uptake of subterranean clover from phosphorus sources of different solubilities. *New Phytologist*, **84**, 327–388.

Parke,J.L., Linderman,R.G. and Black,C.H. (1983) The role of ectomycorrhizas in drought tolerance of Douglas-fir seedlings. *New Phytologist*, **95**, 83–95.

Pearson,V. and Read,D.J. (1973) The biology of mycorrhiza in the Ericaceae, 1. The isolation of the endophyte and synthesis of mycorrhizas in aseptic culture. *New Phytologist*, **72**, 371–379.

Powell,C.Ll. (1984) Field inoculation with VA mycorrhizal fungi. In Powell,C.L1. and Bagyaraj,D.J. (eds), *VA Mycorrhiza*. CRC Press, Boca Raton, pp. 205–222.

Powell,C.Ll. and Bagyaraj,D.J. (eds) (1984) *VA Mycorrhiza*. CRC Press, Boca Raton.

Powell,C.Ll. and Bates,P.M. (1981) Ericoid mycorrhizas stimulate fruit yield of blueberries. *Horticultural Science*, **16**, 655–656.

Read,D.J. (1984) The structure and function of the vegetative mycelium of mycorrhizal roots. In Jennings,D.H. and Rayner,A.D.M. (eds), *The Ecology and Physiology of the Fungal Mycelium*. Cambridge University Press, Cambridge, pp. 215-240.

Read,D.J. and Bajwa,R. (1985) Some nutritional aspects of the biology of ericaceous mycorrhizas. *Proceedings of the Royal Society of Edinburgh, Series B*, **85**, 317–332.

Rhodes,L.H. (1980) The use of mycorrhizae in crop production systems. *Outlook on Agriculture*, **10**, 275–281.

Sanders,F.E. and Sheikh,N.A. (1983) The development of vesicular-arbuscular mycorrhizal infection in plant root systems. *Plant and Soil*, **71**, 223–246.

Sanders,F.E. and Tinker,P.B. (1973) Phosphate flow into mycorrhizal roots. *Pesticide Science*, **4**, 385–395.

Schüepp,H., Dehn,B. and Sticher,H. (1987) Interaktionen zwischen VA-Mykorrhizen und Schwermetallbelastungen. *Angewandte Botanik*, **61**, 85–96.

Singh,J.P., Karamanos,R.E. and Stewart,J.W.B. (1986) Phosphorus-induced zinc deficiency in wheat on residual phosphorus plots. *Agronomy Journal*, **78**, 668–675.

Smith,G.S. (1988) The role of phosphorus nutrition in interactions of vesicular-arbuscular mycorrhizal fungi with soilborne nematodes and fungi. *Phytopathology*, **78**, 371–374.

Smith,S.E. and Gianinazzi-Pearson,V. (1988) Physiological interactions between symbionts in vesicular-arbuscular mycorrhizal plants. *Annual Review of Plant Physiology and Plant Molecular Biology*, **39**, 221–244.

Smith,S.E., St. John,B.J., Smith,F.A. and Bromley,J.-L. (1986) Effects of mycorrhizal infection on plant growth, nitrogen and phosphorus nutrition in glasshouse-grown *Allium cepa* L. *New Phytologist*, **103**, 359–373.

Snellgrove,R.C. and Stribley,D.P. (1986) Effects of pre-inoculation with a vesicular-arbuscular mycorrhizal fungus on growth of onions transplanted to the field as multi-seeded peat modules. *Plant and Soil*, **92**, 387–397.

Spitko,R.A. and Manning,W.J. (1981) Irradiated digested sewage sludge: effects on plant-symbiont associations in the field. *Environmental Pollution (Series A)*, 25, 1–8.

Stribley,D.P. (1987) Mineral nutrition. In Safir,G.R. (ed.), *Ecophysiology of VA Mycorrhizal Plants*. CRC Press, Boca Raton, pp. 59–70.

Stribley,D.P., Tinker,P.B. and Rayner,J.H. (1980) Relation of internal phosphorus concentration and plant weight in plants infected by vesicular-arbuscular mycorrhizas. *New Phytologist*, **86**, 261–266.

Sylvia,D.M., Hung,L.L. and Graham,J.H. (eds) (1988) *Mycorrhizae in the Next Decade: Practical Applications and Research Priorities* (Proceedings of the 7th North American Conference on Mycorrhizae). University of Florida, Gainesville.

Tinker,P.B. (1985) Crop nutrients: control and efficiency of use. *Philosophical Transactions of the Royal Society of London, Series B,* **310**, 175–191.

Tinker,P.B. (1986) Trace elements in arable agriculture. *Journal of Soil Science,* **37**, 587–601.

Tinker,P.B. and Gildon,A. (1983) Mycorrhizal fungi and ion uptake. In Robb,D.A. and Pierpoint,W.S. (eds), *Metal and Micronutrients: Uptake and Utilisation by Plants*. Academic Press, London, pp. 21–32.

Trappe,J.M. (1987) Phylogenetic and ecologic aspects of mycotrophy in the angiosperm from an evolutionary standpoint. In Safir,G.R. (ed.), *Ecophysiology of VA Mycorrhizal Plants*. CRC Press, Boca Raton, pp. 5–25.

Wang,G., Stribley,D.P. and Tinker,P.B. (1985) Soil pH and vesicular-arbuscular mycorrhizas. In Fitter,A.H. (ed.), *Ecological Interactions in Soil*. Blackwell Scientific, Oxford, pp. 219–224.

Woolhouse,H.W. (1988) Agricultural research and agricultural surpluses. *Plants Today*, March–April 1988, 40–42.

CHAPTER 5

The use of microbial inoculants in the biological control of plant diseases

R.CAMPBELL

Department of Botany, University of Bristol, Bristol BS8 1UG, UK

Introduction

Though this review will concentrate on the microbial inoculants, biological control of plant pathogens is much broader in concept and may involve the use of crop management and tillage systems, control of the environment if possible, and sometimes an integration of biological and chemical methods. Indeed, any method of disease control or method of reducing pathogen inoculum brought about in part or entirely by another organism other than man is now regarded as biological control (Baker, 1987). The fact that microorganisms interact with each other, and may inhibit growth or cause death, has been known in culture for more than 100 years (reviewed in Baker, 1987). Actual control of pathogens in systems approximating to natural conditions was demonstrated by Hartley (1921), Millard and Taylor (1927) and by Henry (1931) who used microorganisms in mixtures or from pure cultures against various soil-borne diseases.

Sixty years ago the basic ideas on the use of added inoculants in biological control were established, but until the 1960s these remained largely the province of experimental science rather than field work, and were anyway overshadowed by the enormous development of chemical pesticides during and immediately after World War II.

In 1963, after various preliminary studies, Rishbeth demonstrated the control of stem and butt rot of conifers caused by *Fomes annosus* (now *Heterobasidion annosum*) by inoculating with another fungus, *Peniophora gigantea*. The latter was one of the first organisms developed as a biocontrol agent for commercial use against a plant disease, and it remains in use today.

Since that time there has been an enormous increase in the amount of research effort devoted to biological control (Baker, 1987). This has resulted in more commercial products (such as the fungus *Trichoderma* used against many diseases) and many potential antagonists that are nearing commercial exploitation (of which *Pseudomonas* is the most developed). The increase in interest by government agencies and commercial companies is partly the result of public pressure on environmental protection, especially the desire to reduce the use of pesticides. It is also a result of what is now possible through the increased knowledge of microbial ecology and the opportunities offered by developing techniques, such as genetic engineering and immunological methods of identifying and tracking strains of microbes released into natural or agricultural environments (Andrews, 1986; Hagedorn, 1986).

Current successes of biological control with introduced antagonists

We are concerned here only with those examples where there is either a commercial product or clear evidence that research and development has progressed to a stage where commercial exploitation is a serious probability. There are many thousands of reports of control of diseases in the research literature, which have no possibility of becoming commercial. This may be, of course, because the organism does not work, especially in the field. There are also a number of potentially useful antagonists which have not been, and will not be, developed, although they work well: the crop or disease may not be important enough to warrant development or the information may have been published, so preventing the protection of the producers investment by patents.

Table 1 lists the main antagonists under commercial consideration: it is by no means exhaustive. Most of the examples are for soil-borne diseases rather than foliar pathogens. The main reason for this is that foliar pathogens are often adequately controlled by cheap, effective, chemical methods (usually fungicides) or by varietal resistance with which biological control would have to 'compete'. There are few fungicides effective for soil-borne diseases, especially in the field rather than horticulture, and varietal resistance may not exist (for example against take-all) or may not have been selected for because of the difficulty of working with root diseases. Foliar diseases have naturally been the first target for control because they were the main limitation on plant productivity and they can be relatively easily recognized and assessed in field trials. Root diseases are now seen as a major constraint on productivity (Cook, 1987) and their control is receiving more attention. Biological control on a major scale is a good possibility and is being actively developed. It will be difficult on a field scale because of the long life of the crops and the possibility of repeated attacks by the pathogen, but it is a goal worth aiming for.

Several of the examples in *Table 1* are also specialized in terms of the type of habitat into which an antagonist is introduced. It is very difficult to introduce a 'foreign' organism into an existing community, so the antagonists should naturally grow in the community they are to be used in, as is the case with the *Pseudomonas* spp. which can be vigorous root colonizers (Weller, 1983, 1984). Alternatively, the existing community can be reduced or removed, by the use of fungicides or more general biocides, and the antagonist introduced into the 'biological vacuum' so created. This reduces competition between the existing community and the introduced antagonists. The final solution to the problem is to avoid it, and only use the antagonist in situations where there is little competition. These include virtually sterile nursery composts (*Trichoderma* for damping-off), and the protection of clean wound sites (*Eutypa* on apricot or *Heterobasidion* on conifers). *Peniophora* fails to gain possession of the stump and exclude *Heterobasidion* unless it is inoculated immediately after cutting, when the stump is still a vacant habitat. The exploitation of these situations of low competition has been a major factor in the success of these antagonists, though the creation of the space may well be done, in integrated programs, by chemical means.

A further factor in the success of these antagonists is that they are often designed to give protection for rather short periods of time, on a freshly cut wound surface, on a flower in the case of fire-blight control, or in a seed or seedling where survival for a week or two puts the plant beyond danger from the damping-off pathogens. It

Table 1. Antagonists in, or near to, commercial use.

Antagonist	Pathogen and disease	Comment
Agrobacterium Strain 84	*Agrobacterium tumefaciens* crown gall in horticulture	Special interest in the Ti plasmid as a gene vector from bacterium to eukaryotic host. Antagonist used as a root or cutting dip. Operates by competition for adsorption sites and production of bacteriocin. Biocontrol commercially available (Clare *et al.*, 1987)
Peniophora gigantea	*Heterobasidion annosum*, butt rot of conifers	Antagonist inoculated on to a nearly sterile resource, the cut stem. The first commercially available inoculant. Operates by competitive exclusion and lysis of hyphae (Risbeth, 1975)
Trichoderma spp.	*Pythium, Phytophthora* etc., damping-off of seedlings	Antagonist used as a seed coat or pellet mixed with sterile compost, commercially available. Possibly with fungicide and a fungicide-resistant antagonist. Produces antibiotics, mycoparasite (Chet, 1987; Papavizas, 1985)
	Rhizoctonia, Sclerotinia, Sclerotium etc., various rots of stems, bulbs, etc.	Applied as seed coat or to growing medium, especially to horticultural crop (Papavizas, 1985)
	Chondrostereum purpureum, silver leaf of fruit trees	Antagonist inoculated into stem and reduces existing symptoms and future infection. Commercial trials (Corke and Risbeth, 1981)
Erwinia herbicola	*Erwinia amylovora*, fire blight	Short period of control is needed during flowering. See also *Pseudomonas*. Probably competition (Lindow, 1985)
Fusarium lateritium	*Eutypa armeniacae*, apricot canker	Uses combined fungicide treatment and a fungicide tolerant biocontrol agent. Commercial (Carter, 1983)
Pseudomonas syringae	*Erwinia amylovora*, fire blight	Used in conjunction with control of frost damage. Antibiotics and siderophores probably not important (Beer *et al.*, 1984)
Pseudomonas spp., especially *Ps. putida* and *Ps. fluorescens*	Many diseases, e.g. take-all (*Gaeumannomyces graminis*), *Pythium* root rot, *Rhizoctonia, Fusarium*, etc., also for the control of frost damage	Fluorescent pseudomonads are one of the main groups of bacteria being developed for commercial use. None is yet available, but they are expected on the market by 1989. Produce antibiotics, siderophores and possibly lytic enzymes. (No general reference on *Pseudomonas*: see Cook and Baker, 1983; and for example Weller, 1984)

is relatively easy to have an antagonist survive for a week or two; it is much more difficult to make it last in effective numbers for a whole growing season, although this can be done (Weller, 1983).

Siderophores, biological control of disease and plant growth promotion

Siderophores are chelating compounds with a special affinity for iron, which allow the acquisition of Fe^{3+} and its transport into the cell in iron-limiting environments. Many agricultural soils in England are iron-limited because of the limestone or chalk bedrock or because of the addition of lime to raise the pH, as a result of which most of the iron is in the form of a virtually insoluble hydroxide. Siderophores have received considerable attention as a possible mode of action for biocontrol agents (Swinburne, 1986), though they are also produced by the host plant and frequently by the pathogen, in which case it is not the possession of siderophores which matters, but their relative affinity for iron. The situation is complicated by diseases where siderophore production by an organism other than the pathogen or host makes the disease worse, not better.

There are examples, however, usually with *Pseudomonas* spp., where siderophore production seems to be a major part in the biological control, depriving the pathogen of iron and making it grow more slowly or not at all. In *Fusarium oxysporum* f.sp. *cucumerinum* and f.sp. *lini*, the *Pseudomonas* siderophore inhibits chlamydospore germination and hence reduces disease (Baker *et al.*, 1986). The hosts produce siderophores which have a slightly greater affinity for iron than those from *Fusarium*, but the *Ps. putida* siderophore has a greater affinity than both, so except under acid conditions where iron may be in excess, the *Ps. putida* deprives the *Fusarium* of iron and reduces disease. There are various chemical chelators which have intermediate affinities and can be used to control iron levels in soil artificially and these show that if the *Fusarium* is deprived of iron, either artificially, by a host or by *Ps. putida*, then disease is reduced (Scher and Baker, 1982). Other work shows that it is possible to produce disease reduction with pure siderophore (the chemical without the bacterium), though the possibility remains that the mechanism is an antibiotic effect rather than reduction of iron concentration. Some biocontrol pseudomonads produce siderophores and antibiotics and both seem to be necessary for the full action of the organism (Weller and Cook, 1986).

Pseudomonads and siderophores have also involved so-called plant growth promotion. Plant-growth-promoting rhizobacteria (PGPR) can be inoculated on to some crops and can give increased growth. Many bacteria can produce plant growth hormones in culture, though proof of such activity under normal field conditions does not exist (Beringer *et al.*, Chapter 8). Microbes can also affect mineral nutrition of plants. There are therefore possibilities for altering plant growth which are distinct from any biological control of diseases.

The effectiveness of PGPR after inoculation can often be demonstrated only in normal soil or in soil after a particular crop sequence. Here it is probable that the inoculated bacteria are controlling minor root pathogens which are not producing enough obvious symptoms to be noticed. Elimination of this hidden disease, by the supposed PGPR (or by soil fumigation), causes an apparent growth response, though this is not promoting growth *per se* so much as removing a harmful effect of a pathogen and allowing a more complete expression of the plant's potential.

A similar effect is shown by potatoes which, when grown frequently in the same ground, may give reduced yields. Addition of so-called PGPR restores the situation to near normal. The PGPR have no effect on healthy potatoes in soil which has not been used for this crop for at least some years (Geels and Schippers, 1983). Again it is controlling a deleterious effect, not promoting growth of the normal crop. In this case bacteria which grow in the rhizosphere and produce cyanide are thought to be involved. The ability to produce cyanide is quite common in bacteria and may result from microbial β-glucosidase activity on plant-produced cyanogenic glucosides (Dartnall and Burns, 1987) or from the actual production of cyanide by the harmful bacterium. The cyanide-producing bacteria increase in numbers during repeated cropping of potatoes, and the cyanide is thought to reduce root growth and metabolism and be responsible for the poor yield of the plant. The PGPR which control this situation produce a siderophore which deprives the harmful bacteria of iron, reduces the cyanide production and allows the potato to grow better (Schippers et al., 1987). These results have been confirmed by showing that Tn5-transposon mutants, which lack the ability to produce siderophores, do not show any effect. The field trials were, however, somewhat variable, and the organisms are not always effective (Geels et al., 1986). This is common to many biological control situations (see below).

There is therefore good evidence that siderophores are involved in some forms of biological control, especially where the antagonists are pseudomonads, but their action is not always conclusive and there are many antagonists which do not produce them or in which they do not seem to be important. Siderophores are only one of many different modes of action known.

Minor pathogenic effects and problems with continuous cropping are widely known to farmers. Other replant problems occur, especially with some orchard crops which have a very long life: it is difficult to establish a good crop on land previously growing the same sort of tree. This is most likely to be caused by the build-up of pathogens or other deleterious organisms, and it may be possible to use biological inocula to improve the performance of later crops, though the problems of controlling such long-term infections could be considerable.

Competition for nutrients

Competition occurs between organisms for nutrients and available useful space (space where conditions are suitable for growth). It is already used in biocontrol (for example with *Peniophora, Table 1*), and several systems employing it are under serious development.

The best known of these consists of the ice nucleation bacteria (Lindow, 1985; Beringer et al., Chapter 8). Ice damage is caused to plants when ice crystals form, withdrawing water from the cells and causing physical damage. Crystals will not form until water reaches quite low temperatures ($-5\,°C$ or even $-10\,°C$) unless there are ice-nucleation sites. These may be other crystals or particles of dust; certain proteins, produced by *Pseudomonas*, can serve as ice nuclei and ice damage is more severe in the presence of these bacteria. Lindow's group has removed the gene for this protein and the ice-minus (ice$^-$) mutants compete successfully with the normal population on leaves and flowers so that there are fewer ice nuclei and less damage. Experiments have shown that antibiotics and siderophores are not important in this interaction. There have been problems with the release of this genetically engineered organism for field trials, but the system looks most promising and will surely be developed in some form.

The other good example of competition in biological control is in the study of

Fusarium-suppressive soils of the Chateaurenard region of France. These soils do not permit the development of *Fusarium oxysporum* wilt, especially on melons, though the pathogen is present. Extensive work has shown that the control is given by other species of *Fusarium* or other *formae speciales* of *F.oxysporum* (Alabouvette et al., 1986) and the main interaction is competition for carbon, especially during spore germination. There is evidence that competition for iron, via *Fusarium* siderophore systems, may also be involved in this suppression of wilt (Lemanceau et al., 1986). This control of wilt is transferable between soils, so the possibility exists of using it on a wider scale, though it clearly depends on competition for available niches between closely related organisms and may not be widely applicable. Alternatively, the competing fungi have been isolated and could be used as inocula.

There are many other examples of suppressive soils known, and they have been a useful source of potential antagonists, for much of the suppressiveness is of microbial origin. Not all the microorganisms isolated from suppressive soils work by competition.

Antibiotic producers as biocontrol agents

There are many screening procedures which select potential biocontrol agents on the basis of antibiotic inhibition zones on agar plates. The fact remains, however, that antibiotic production in realistic concentrations has only rarely been shown in normal soil. Some biocontrol agents may work, at least partly, by antibiotic production. This is known for the control of take-all disease of wheat (caused by *Gaeumannomyces graminis*) where fluorescent pseudomonads, isolated from decline soils (see below), operate partly by the use of antibiotics (Weller and Cook, 1986; Cook, 1988). The possibility also exists of using such organism to produce the effective antibiotic in industrial fermentation systems and using the chemical itself as a control agent. This may be the way forward for antibiotic producers, though whether it is still biological control is doubtful.

The phenomenon of take-all decline, mentioned above, occurs when successive crops of wheat or other cereals are grown. After a few years the disease naturally becomes less severe (the take-all disease declines) and this is thought to be associated with changes in the soil microbiology. Decline soils, which are also known for other diseases, are a special case of suppressive soils, and again they have been used as a source of potential biological control agents which have subsequently been used as microbial inoculants.

A specialized case of biocontrol with an introduced antagonist is crown gall caused by *Agrobacterium tumefaciens* (Clare et al., 1987; Thomson, 1987; Beringer et al., Chapter 8). The antagonist (strain 84) is available commercially (*Table 1*) and it lacks the Ti-plasmid, which induces the production of growth hormones in the host, and is therefore avirulent. It can compete with the pathogen for attachment sites on the host, and it also produces an antibiotic (a bacteriocin, agrocin 84) to which many strains of *A.tumefaciens* are sensitive. Agrocin 84 uses the same membrane transport system as some opines, which the host produces when infected with the Ti-plasmid and which *A.tumefaciens* can catabolize.

Mycoparasites and biological control

Some of the best-known fungal antagonists kill pathogens by parasitism. *Trichoderma* has been especially studied by Chet and co-workers in Israel (Chet, 1987) who have shown that *T.harzianum* can control some *Pythium* spp. by mycoparasitism, though it is clear that these and other *Trichoderma* spp. also have other modes of action, such as competition. *Sporidesmium* and *Coniothyrium* attack sclerotia, especially of *Sclerotium cepivorum*, *S.sclerotiorum* and *Sclerotinia minor* (Ayers and Adams, 1979, 1981). Many other mycoparasites are known.

Mycoparasites need the host, the plant pathogen, to be present in the soil or on the plant, and they may therefore not be best as protectants. They could, however, be very successful at reducing inoculum in the soil, with or without the host, especially if this was in the form of sclerotia, which represent a considerable food base for the mycoparasite. Some of these pathogens are a particular problem because of the longevity of the sclerotia and the limited number of chemicals (such as pentachloronitrobenzene) available for field use. So, provided some low level of disease can be tolerated, mycoparasites have promise for reducing inoculum potential. Drastic chemical treatments like fumigation with methyl bromide give control of *S.rolfsii* and *Rhizoctonia solani*, but re-infestation can be rapid. Combined fumigation and introduction of *T.harzianum* can give very good and lasting control, and this integrated program shows more promise than either chemical or biological methods alone (Elad *et al.*, 1982).

The prospects of biological control using introduced inoculum

This possibility is of necessity speculative, but it could be useful to try to focus on likely systems for development. We are overwhelmed at the moment by many laboratory examples and possibilities, few of which are ever going to be employed on a wide scale. The problem is how we choose the ones to develop.

The production of commercially useful biological control agents is a long process and needs research and development time and money. Most of those which are now commercial or near-commercial have been worked on, with varying intensity, for about 20 years. There is no reason *a priori* why the development of a biological control agent should be any less expensive than a chemical one though testing and registration may be cheaper (Lethbridge, Chapter 2). Detailed cost analysis and risk assessments have not been published for biocontrol agents of plant diseases, though they are likely to be rather different from chemicals if they follow the assessment for biological control of pests (Reichelderfer, 1981). To fund such research the potential market will need to be considerable, so we need to look for major crops of economic importance. Biological disease control in the few really important field crops (cereals, potatoes, soyabean, vines) is being studied: some work is done on rice diseases but this may be economically less important, as rice is mostly grown in the developing countries where markets for inoculants may be small (though other means of biological control are relevant). High-value horticultural crops are good targets for biocontrol inoculants, and horticulture also allows some control of environment and sometimes the provision of sterile composts into which it is easier to introduce antagonists. Other high-value, intensively controlled systems, such as mushroom growing, are also possibilities, and biological control of some diseases, like bacterial brown blotch, is being developed. The other situations where there is little competition for the antagonist, such as tree stems and wounds (*Table 1*), are also worth considering seriously for further development.

As regards the sorts of diseases which should be targeted for biological control, the present successes and development projects are on roots rather than leaf diseases, for the reasons outlined above.

The sort of antagonist to investigate or to search for is difficult to decide. It is likely that specific organisms or strains are needed for particular diseases, and it is always worth looking for new and better organisms (Campbell, 1986). It is, however, also clear that past experience suggests that *Pseudomonas* and *Trichoderma* are particularly important genera and might be especially considered for any new disease-control situation. It is also necessary to select an organism suitable for the environment: colonization ability is important on roots, and ruderal organisms may be preferable to colonize and exploit rapidly the environment to which they are introduced.

The overall approach to the disease must also be considered, as well as the addition of antagonists. Changes may be needed in cropping sequences and agricultural practices. The introduced antagonist stands a much better chance of survival and success if the balance of environmental conditions can be tipped a little in its favour. There is evidence (Neal *et al.*, 1970) that the host genome controls rhizosphere populations to some extent, and there may be varietal differences in the colonization or effectiveness of antagonists which could be exploited. In the future, plant breeding and genetic engineering may 'design' the host and antagonist to be compatible and mutually harmful to the pathogen. Agricultural practices such as green manuring and undersowing cereals with legumes can reduce disease, probably by changing the microflora, and there is also some evidence that the effects of applied antagonists may be enhanced by such treatments (M.Lennartsson, unpublished). We need to consider antagonists in terms of the overall crop system (Cook, 1988). In the more industrialized agricultural systems this includes integrating biological and chemical control. There are already successful examples of this (*Table 1*), and it should not be looked upon as a problem with biological control, but rather as a technique which can be used to give the biocontrol agent a vacant niche and a competitive advantage over the pathogen and less useful saprophytes.

There remain, however, some obvious problems with biological control, the most topical of which is the worry over release of selected and genetically engineered organisms (Panopoulos, 1986; Beringer *et al.*, Chapter 8). These will however be overcome when suitable protocols and rules for testing and release have been agreed. Much more fundamental to biological control, including plant growth promotion by minor pathogen control, is the problem of repeatable results. Many claims have been made over the years to have controlled a disease, but many have come to nothing becaue the next year the experiments failed. Organisms may work in one field and fail in another in the same year (Capper and Campbell, 1986; Geels *et al.*, 1986). There are many reasons for this. Microbial antagonists are alive and growing, and hence more variable and more subject to environmental effects than a chemical from a bottle. Some microbes do not work in some soil types or at some water potentials, and so on. These problems may be solved by finding organisms (like the pseudomonads) that are nutritionally and physiologically very adaptable, or by using a mixture of organisms with different characteristics, although antagonists can antagonize each other and may not work better in mixtures. There is also a serious possibility of using an ineffective organism: a microbe is selected and tested in the laboratory and the field, fermenter conditions are designed for its mass production, it goes through safety testing and many other investigations, but great care is needed to ensure that the organisms finally applied as an inoculant are those that have been so carefully tested and shown to work (see Lethbridge, Chapter 2). A laboratory- and fermenter-adapted strain which will not grow in the field must not be produced.

There is also a problem with soil-borne diseases, which are the ones mainly worked on, in ensuring your field sites have adequate levels of disease to test the potential biocontrol agent. Soil-borne diseases are often patchy in distribution, difficult to forecast from year to year, or difficult to assess. It can be hard to judge whether treatment of any sort is needed. Furthermore, distinction must be made between the failure of the potential biocontrol agent to control the disease, and the failure to test the organism against a reasonable level of pathogen. In experimental terms, it is necessary to know whether a lack of statistical difference between the treatments and the control was because the treatments were ineffective or whether the controls and treatment plots were both healthy and so could not be improved upon.

For all these reasons, it is common for only half of the trials, or even less, to show the introduced organisms to be effective (Geels *et al.*, 1986; Suslow, 1982). This rate must be improved for commercial use and should be one of the main concerns in future research.

Biological control by introduced inocula is already in use for the control of some diseases in intensive agriculture. It will become increasingly important, especially in terms of direct manipulation of organisms, in the future, and will run alongside chemical treatments. In less intensive, low-input agricultural systems, in developing countries for example, biocontrol of plant diseases is likely to be even more important, since chemical methods may be too expensive. Methods based on high-technology fermentation systems and the mass introduction of antagonists are likely to be less important in these situations than the manipulation of existing organisms by agricultural practices. However, the production and use of *Rhizobium* (Eaglesham, Chapter 3) has shown that even mass introduction can have a role in less intensive agriculture.

References

Alabouvette,C., Coutendier,Y. and Lemanceau,P. (1986) Nature of intrageneric competition between pathogenic and non-pathogenic *Fusarium* in a wilt-suppressive soil. In Swinburne,T.R. (ed.), *Iron Siderophores and Plant Diseases* (NATO AS1 Series no. 117). Plenum Press, New York, pp. 165–178.

Andrews,J.H. (1986) How to track a microbe. In Fokkema,N.J. and van den Heuvel,J. (eds), *Microbiology of the Phyllosphere*. Cambridge University Press, Cambridge, pp. 14–34.

Ayers,W.A. and Adams,P.B. (1979) Mycoparasitism of sclerotia of *Sclerotinia* and *Sclerotium* species by *Sporidesmium sclerotivorum*. *Canadian Journal of Microbiology*, **25**, 17–23.

Ayers,W.A. and Adams,P.B. (1981) Mycoparasitism and its application to biological control of plant diseases. In Papavizas,G.C. (ed.), *Biological Control in Crop Production*. Allanheld, Osmum Publishers, Granada, pp. 91–105.

Baker,K.F. (1987) Evolving concepts of biological control of plant pathogens. *Annual Review of Phytopathology*, **25**, 67–85.

Baker,R., Elad,Y. and Sneh,B. (1986) Physical, biological and host factors in iron competition in soils. In Swinburne,T.R. (ed.), *Iron Siderophores and Plant Diseases* (NATO AS1 Series no. 117). Plenum Press, New York, pp. 77–84.

Beer,S.V., Rundle,J.R.and Norelli,J.L. (1984) Recent progress in the development of biological control for fire blight—a review. *Acta Horticulturae*, **151**, 195–201.

Campbell,R. (1986) The search for biological control agents, a pragmatic approach. *Biological Agriculture and Horticulture*, **3**, 317–327.

Capper,A.L. and Campbell,R. (1986) The effect of artificially inoculated antagonistic bacteria on the prevalence of take-all disease of wheat in field experiments. *Journal of Applied Bacteriology,* **60**, 155–160.

Carter,M.V. (1983) Biological control of *Eutypa armeniacae*. 5. Guidelines for establishing routine wound protection in commercial apricot orchards. *Australian Journal of Experimental Agriculture and Animal Husbandry,* **23**, 429–436.

Chet,I. (1987) *Trichoderma*—application, modes of action and potential as a biocontrol agent of soilborne plant pathogenic fungi. In Chet,I. (ed.), *Innovative Approaches to Plant Disease Control*. John Wiley, New York, pp. 137–160.

Clare,B.G., Petit,A. and Tempé,J. (1987) The biology of pathogenic plasmids of *Agrobacterium*. In Day,P.R. and Jellis,G.J. (eds), *Genetics and Plant Pathogenesis*. Blackwell Scientific, Oxford, pp. 79–90.

Cook,R.J. (1987) Pathogens as constraints to crop productivity. In Jordan,W.R. (ed.), *Water and Water Policy in World Food Supplies*. Texas A & M University Press, College Station, pp. 229–249.

Cook,R.J. (1988) Management of the environment for the control of pathogens. *Philosophical Transactions of the Royal Society, Series, B,* **318**, 171–182.

Cook,R.J. and Baker,K.F. (1983) *The Nature and Practice of Biological Control of Plant Pathogens*. American Phytopathological Society, St Paul, p. 539.

Corke,A.T.K. and Rishbeth,J. (1981) Use of micro-organisms to control plant diseases. In Burges,H.D. (ed.), *Microbial Control of Plant Pests and Diseases*. Academic Press, London, pp. 717–736.

Dartnall,A.M. and Burns,R.G. (1987) A sensitive method for measuring cyanide and cyanogenic glucosides in sand culture and soil. *Biology and Fertility of Soils,* **5**, 141–147.

Elad,Y., Hadar,Y., Chet,I. and Henis,Y. (1982) Prevention, with *Trichoderma harzianum* Rifai aggr., of reinfestation by *Sclerotium rolfsii* Sacc. and *Rhizoctonia solani* Kuhn of soil fumigated with methyl bromide, and improvement of disease control in tomatoes and peanuts. *Crop Protection,* **1**, 199–211.

Geels,F.P. and Schippers,B. (1983) Reduction in yield depressions in high frequency potato cropping soil after seed tuber treatments with antagonistic fluorescent *Pseudomonas* spp. *Phytopathologische Zeitschrift,* **108**, 207–214.

Geels,F.P., Lamers,J.,G., Hoekstra,O. and Schippers,B. (1986) Potato plant response to seed tuber bacterization in the field in various rotations. *Netherlands Journal of Plant Pathology,* **29**, 257–272.

Hagedorn,C. (1986) Role of genetic variants in autecological research. In Tate,R.L. (ed.), *Microbial Autecology*. John Wiley, New York, pp. 61–73.

Hartley,C. (1921) Damping-off in forest nurseries. *US Department of Agriculture Bulletin,* **934**, 1–99.

Henry,A.W. (1931) The natural microflora of the soil in relation to the foot rot problems of wheat. *Canadian Journal of Research,* **4**, 69–77.

Lemanceau,P., Alabouvette,C. and Meyer,J.M. (1986) Production of fusarinine and iron assimilation by pathogenic and non-pathogenic *Fusarium*. In Swinburne,T.R. (ed.), *Iron Siderophores and Plant Diseases* (NATO AS1 Series No. 117), Plenum Press, New York, pp. 251–260.

Lindow,S.E. (1985) Integrated control and the role of antibiosis in biological control of fireblight and frost injury. In Windels,C.E. and Lindow,S.E. (eds), *Biological Control on the Phylloplane*. American Phytopathological Society, St Paul, pp. 83–115.

Millard,W.A. and Taylor,C.B. (1927) Antagonism of microorganisms as the controlling factor in the inhibition of scab by green manuring. *Annals of Applied Biology,* **14**, 202–216.

Neal,J.L., Atkinson,T.G. and Larson,R.I. (1970) Changes in the rhizosphere microflora of spring wheat induced by disomic substitution of a chromosome. *Canadian Journal of Microbiology,* **16**, 153–158.

Panopoulos,N.J. (1986) Tactics and feasibility of genetic engineering of biocontrol agents. In Fokkema,N.J. and van den Heuvel,J. (eds), *Microbiology of the Phyllosphere*. Cambridge University Press, Cambridge, pp. 312–332.

Papavizas,G.C. (1985) *Trichoderma* and *Gliocladium*: biology, ecology and potential for biological control. *Annual Review of Phytopathology,* **23**, 23–54.

Reichelderfer,K.H. (1981) Economic feasibility of biological control of crop pests. In Papavizas,G.C. (ed.), *Biological Control in Crop Production*. Allanheld, Osmum Publishers, Granada, pp. 403–417.

Rishbeth,J. (1963) Stump protection against *Fomes annosus*. III. Inoculation with *Peniophora gigantea*. *Annals of Applied Biology,* **52**, 63–77.

Rishbeth,J. (1975) Stump inoculation: a biological control of *Fomes annosus*. In Bruehl,G.W. (ed.), *Biology and Control of Soil-borne Plant Pathogens*. American Phytopathological Society, St Paul, pp. 158–172.

Scher,F.M. and Baker,R. (1982) Effect of *Pseudomonas putida* and a synthetic iron chelator on induction of soil suppressiveness to *Fusarium* wilt pathogens. *Phytopathology,* **72**, 1567–1573.

Schippers,B., Bakker,A.W. and Bakker,P.H.A.M. (1987) Interactions of deleterious and beneficial rhizosphere microorganisms and the effect of cropping practices. *Annual Review of Phytopathology,* **25**, 339–358.

Swinburne,T.R. (ed.) (1986) *Iron Siderophores and Plant Diseases* (NATO AS1 Series No. 117). Plenum Press, New York.

Suslow,T.V. (1982) Role of root-colonizing bacteria in plant growth. In Mount,M.S. and Lacey,G.H. *Phytopathogenic Prokaryotes.* Vol. 1. Academic Press, New York, pp. 187–224.

Thomson,J.A. (1987) The use of agrocin-producing bacteria in the biological control of crown gall, In Chet,I. (ed.), *Innovative Approaches to Plant Disease Control.* John Wiley, New York, pp. 213–228.

Weller,D.M. (1983) Colonization of wheat roots by a fluorescent pseudomonad suppressive to take-all. *Phytopathology,* **73**, 1548–1553.

Weller,D.M. (1984) Distribution of a take-all suppressive strain of *Pseudomonas fluorescens* on seminal roots of wheat. *Applied and Environmental Microbiology,* **48**, 897–899.

Weller,D.M. and Cook,R.J. (1986) Suppression of root disease of wheat by fluorescent pseudomonads and mechanisms of action. In Swinburne,T.R. (ed.), *Iron Siderophores and Plant Diseases* (NATO AS1 Series No. 117), Plenum Press, New York, pp. 99–107.

CHAPTER 6

The use of plant viruses as inoculants

R.F.WHITE AND J.F.ANTONIW

Plant Pathology Department, AFRC Institute of Arable Crop Research, Rothamsted Experimental Station, Harpenden, Herts AL5 2JQ, UK

Plant viruses can be used as inoculants for two purposes, to make plants resistant to other pathogens, but especially viruses, and to give plants desirable characteristics other than resistance to pathogens.

Inoculants to make plants resistant to other pathogens, but especially viruses

Nearly 60 years ago, McKinney (1929) found that plants infected with one strain of tobacco mosaic virus (TMV) could not be infected by another strain. Since then there have been many studies using both closely related and unrelated viruses in investigations of this phenomenon. In this chapter we shall only describe work where the plant was inoculated with the protecting virus before the challenge virus.

Most plant viruses induce symptoms in their hosts, and the practical aim of using a virus which causes no symptoms as an inoculant is to reduce or eliminate the symptoms should a 'severe' virus subsequently infect the plant. The symptomless virus does not necessarily have to make the plant immune to subsequent infection to prevent the challenging virus from causing symptoms.

The great majority of plant viruses consist of nucleic acid (usually RNA) protected in a protein coat which is coded for by the virus nucleic acid. To multiply, the virus must enter the plant cell and utilize the nucleic acid and protein synthesizing machinery of the host. This disruption of normal cellular metabolism produces symptoms in the plant either directly or indirectly.

Usually when a plant is infected with a virus, the virus either multiplies and spreads systemically throughout the plant, or the plant will respond to infection and localizes the virus to a few cells surrounding the initial point of infection. The role and use of inoculants for the induction of resistance is determined by these reactions.

Protection induced in plants systemically infected with a virus

This phenomenon has often been described as 'cross-protection', and there are examples of its successful exploitation (Fletcher and Rowe, 1975; Costa and Muller, 1980; Po Tien *et al.*, 1987). Nevertheless these successes have been relatively few because of several drawbacks to this type of control. For example, mild symptomless virus strains protect one crop but may spread to another crop and cause serious disease. Some viruses which alone give few symptoms can react synergistically and produce severe symptoms when they are together in the same plant, so care must be taken that such a reaction does not occur between the protecting virus and another unrelated virus common in the crop. Virus infection may make the plant more susceptible to diseases caused by other pathogens (Russell, 1970). Mutants may arise from the protecting strain that give symptoms, or new mutants might come into the crop against which the protecting virus has no effect. Finally, inoculation of the crop can be expensive or impossible, as each plant has to be inoculated mechanically or with a vector or by grafting and therefore high-value perennial or vegetatively propagated crops are the best targets for this type of protection.

Because of these difficulties, cross-protection has only been attempted where the disease was impossible to eradicate, where it could not be adequately controlled and where mild strains existed which caused little or no apparent yield loss. The selection of mild virus strains must be done with care, and only those that protect against all virus strains in all varieties of the crop found in a particular geographical areas should be used. The best strategy is to isolate naturally-occurring mild strains which give protection to individual plants in a heavily infected crop. There have been three relatively successful attempts to use such mild strains for protection in the field, the crops involved being tomato, citrus and pepper.

Tomato mosaic virus (ToMV) was a very difficult virus to control in tomato (*Lycopersicon esculentum*) as it is very easily transmitted by mechanical inoculation, and cultural practices in glasshouses, such as removing side shoots by hand, rapidly spread the virus through the crop. Natural strains which cause mild symptoms were used for crop protection but were superseded by strains produced by nitrous acid mutagenesis. These strains gave good protection against a variety of strains with only very slight symptoms (Rast, 1972) and were used throughout the world, particularly in Europe and Japan, giving increased yields of approximately 7% in the UK (Channon *et al.*, 1978) and 15% in Holland (Broadbent, 1976). In 1974, 70% of tomatoes grown in Japan were inoculated with mild ToMV strains (Fletcher and Rowe, 1975). Methods were devised (for example, high-pressure spray guns containing virus inoculum and an abrasive) for large-scale inoculation of tomato seedlings, but in some cases cross-protection failed, either because plants were not protected because of inefficient inoculation or because the mild strains failed to protect against all strains of the virus (Fletcher and Rowe, 1975). These problems, and the small but significant reduction in yield caused by the mild strains, persuaded most growers to switch to ToMV-resistant varieties of tomato when they became available. Few growers now inoculate their plants with ToMV mild strains.

Citrus tristeza is a severe, aphid transmitted virus disease now found in most citrus-growing areas. Citrus is a perennial, vegetatively propagated crop and is therefore an ideal candidate for inoculation with a mild strain. The mild strains were derived from infected but symptomless trees and in some cases have given good protection against the tristeza disease. In Brazil (Costa and Muller, 1980) over 8 million trees infected with mild strains have largely remained free of the damaging disease. In other regions protection has not been so effective, and some infected trees have given rise to virus strains which, when naturally transmitted, have produced serious disease (Bar-Joseph, 1978).

Cucumber mosaic virus (CMV) infects many agricultural crops and is important in peppers (*Capsicum* spp.). The main method of control is to destroy the aphid vector of the virus but in China and elsewhere this has proved difficult (Po Tien *et al.*, 1987). The virus genome consists of 3 RNA components and there is a subgenomic RNA 4 which codes for the coat protein. Sometimes a small satellite RNA is present which cannot replicate alone and is dependent on CMV for its replication. This satellite RNA considerably reduces the symptoms produced by CMV (Kaper and Tousignant, 1984) in a wide range of plants (Po Tien *et al.*, 1987). A mild strain of CMV containing the satellite RNA has been successfully used to protect pepper plants from infection with CMV in 14 localities in China (Po Tien *et al.*, 1987). The severity of the disease was reduced, and fruit yields were increased by 11−56% depending on the locality. In the field, plants were inoculated using a spray gun containing the virus and its satellite RNA together with an abrasive. Careful preparation of the inoculum is required to make sure that all plants are infected with both CMV and the satellite RNA.

Mechanisms of cross-protection. If we could understand the mechanisms involved with this type of protection, it might be possible to induce it with chemicals or by genetic manipulation rather than with infectious virus. In general, the more closely viruses are related, the greater the amount of cross-protection they give against each other. Experiments with plant protoplasts suggest that if a virus multiplies within a cell, it prevents the challenge virus from replicating (Otsuki and Takebe, 1976; Barker and Harrison, 1978). If the challenge virus does multiply, it has difficulty in spreading (Barker and Harrison, 1978). Experiments using protoplasts must be interpreted with care, however, as the leaf may respond as a 'tissue' rather than a collection of individual cells (Shalla and Peterson, 1978). Cross-protection may be a plant resistance mechanism stimulated by the primary virus infection and is therefore more effective against a challenge infection with a similar virus. Alternatively, viruses may have developed cross-protection in order to prevent mutant viruses from out-competing the parent strain. By whatever means it has evolved, cross-protection involves more than one mechanism, as it is effective against viruses with different replication strategies, and its form differs depending on virus and host.

Several hypotheses have been put forward to explain cross-protection, and some of the more plausible were summarized by Ponz and Bruening (1986). The cross-protection could operate by any one, or any combination, of the following. The nucleic acid of the challenge virus may be made inaccessible by the coat protein of the replicating virus. This occurs either because the challenge virus cannot uncoat in the presence of the replicating virus coat protein or the nucleic acid of the challenge virus is coated by the protein before it can initiate an infection. Sherwood and Fulton (1982) and Dodds *et al.* (1985) found that in some cases virus RNA could infect protected tissue whereas whole virus could not. This suggests that protection is due to the inability of the challenge virus to uncoat.

The RNA of the challenge virus may be prevented from initiating infection by the presence of 'antisense' complimentary RNA of the protecting virus. It has been shown in both bacterial (Pines and Inouye,1986) and eukaryotic cells (Kim and Wold, 1985) that RNA-controlled processes can be inhibited by RNA 'antisense' of the biologically active form. Thus when the challenge virus uncoats, 'antisense' RNA molecules produced during the primary infection hybridize with the challenge virus RNA and prevent its translation and replication.

Plant viruses use host-derived factors to replicate. Some or all of these factors may be blocked or depleted by a primary virus infection, so that the challenge virus is unable to multiply.

The challenge virus may be unable to replicate because either its replicase is insufficiently selective and replicates the virus already present in the cell or the replicase of the protecting virus combines with the challenge virus but is non-productive.

The protecting virus may produce a (virus-encoded) substance that prevents the multiplication of the challenge virus. Some, and maybe all, plant viruses encode factors which appear to be essential for the normal spread of the virus within the plant. Perhaps the factor produced by the protecting virus is not effective for the challenge strain but saturates all the available sites and blocks the action of the challenge virus factor.

Two of these hypotheses have been tested by producing transgenic plants. Baulcombe et al. (1987) found that plants containing antisense RNA to parts of the genome of tobacco rattle virus were as susceptible to the virus as control plants, and the spread of virus within the plant was unaffected. They suggested that because antisense inhibition is not 100% (Kim and Wold, 1985) it would allow the virus to multiply slightly, and the antisense effect would be lost very quickly as more virus was produced.

Powell Abel et al. (1986) and Nelson et al. (1987) used transgenic tobacco plants that produced the coat protein of TMV. They found that these plants had fewer sites available for infection and that, once infection was established, both virus replication in the inoculated leaf and its spread to the upper leaves was reduced. Transgenic tomato plants containing the TMV coat protein took much longer than control plants to develop symptoms when inoculated with TMV.

Tumer et al. (1987) transformed both tobacco and tomato plants to produce the coat protein of alfalfa mosaic virus (AlMV), and found a similar type of resistance to AlMV, as described above. Van Dun et al. (1987) and Bol (1987), however, transformed tobacco plants to produce the coat protein of AlMV and found them to be completely resistant to AlMV. It seems likely, therefore, that at least part of the mechanism of cross-protection in some plants infected with some viruses involves the action of the coat protein of the protecting virus.

The action of coat protein is unlikely to be involved in the reduction of symptoms by satellite virus, and it is not known how this works. However, Harrison et al. (1987) found that plants genetically engineered to produce CMV satellite RNA did not show symptoms when inoculated and infected with CMV, thus mimicking the action of satellite RNA in natural infections. A similar strategy might be used with other virus−satellite−host combinations.

There is enormous potential for producing genetically engineered plants showing resistance to virus infection (Beringer et al., Chapter 8). Most of the drawbacks associated with using virus inoculants listed earlier are removed when a plant is transformed to express only part of the virus genome. The work with coat proteins suggests that these 'partial' inoculants could be extremely important in reducing crop losses due to virus diseases, particularly in areas of the world where present control methods are ineffective.

Protection induced in plants after infection with a virus that is localized

Most plants have the ability to respond to certain viruses by localizing the virus to a few cells around the point of the infection, and a chlorotic or, more commonly, a necrotic lesion is formed. Death of the cells in the lesion, however, is not the reason for virus localization, as virus particles can be found in a small zone of living cells around the lesion. Many viruses infect some host species systemically but are localized in others, and many different hosts have the ability to localize some viruses. This is a very effective form of resistance and, in crops, localized virus infections are of no economic importance.

Localization of the virus affects tissues that have not been infected, making them resistant to a challenge inoculation. Both bean and tobacco leaves with localized TMV infections have very resistant tissue bordering the local lesions (Yarwood, 1960; Ross, 1961a) and this induced resistance is also shown against unrelated localized viruses. The induced resistance can also be found in other leaves on the inoculated plant, and this systemic acquired resistance shows the same form as that in the inoculated leaf: challenge inoculation gives fewer and smaller lesions and the challenge virus does not have to be related to the first infecting virus (Ross, 1961b; Ross, 1966).

This uninfected virus-resistant tissue also shows induced resistance to bacterial and fungal infection and, conversely, bacterial and fungal pathogens which produce localized necrotic lesions induce resistance to viruses (for review see Gianinazzi, 1983). Thus, as in cross-protection, more than one resistance mechanism appears to be involved.

Mechanisms of induced resistance. Attempts to understand the mechanisms involved in induced resistance have centred on the production of new proteins in the uninfected resistant tissue. Loebenstein and co-workers have isolated a protein (26 kD) from resistant tobacco leaf tissue which, when applied to healthy leaves, leaf discs or protoplasts, makes them resistant to virus infection. They were unable to detect this protein, called inhibitor of viral replication (IVR), from systemically infected tissue and found it to be neither host-specific nor virus-specific. IVR has been detected in leaves showing systemic acquired resistance, and may well be the antiviral agent in resistant tissue (Loebenstein, 1987). It will be interesting to see if plants transformed by genetic manipulation to produce IVR constitutively are resistant to virus infection.

In contrast to IVR, which is produced in only small quantities in resistant tissue, pathogenesis-related (PR) proteins (Antoniw *et al.*, 1980) are produced in large amounts in response to localized infection by various pathogens. PR proteins are found in a wide range of plants (Van Loon, 1985; White *et al.*, 1987) and appear to accumulate in the intercellular leaf space (Parent and Asselin, 1984). At least 23 PR proteins are induced in tobacco plants with a localized reaction to TMV (Hogue and Asselin, 1987) and some of these proteins almost certainly have an antibacterial and or antifungal action as they are glucanases and chitinases (Fritig *et al.*, 1987).

There is a very good correlation between PR protein induction and induced resistance to virus infection (White *et al.*, 1986; Antoniw and White, 1986). Several chemicals (Gianinazzi, 1983; White *et al.*, 1986) induce both resistance and PR proteins in the treated leaf, and this chemically-induced resistance is not due to a non-specific stress (White *et al.*, 1986). The interspecific hybrid of *Nicotiana glutinosa* × *N. debneyi* constitutively produces PR proteins and is highly resistant to virus infection (Ahl and Gianinazzi, 1982), suggesting that PR proteins are closely associated with, but not necessarily responsible for, induced resistance. This interspecific hybrid is, however, susceptible to infection with systemic viruses as are plants showing induced resistance after chemical treatment (Gianinazzi,S., Ahl,P., Antoniw,J.F. and White,R.F., unpublished) or inoculation with localized viruses (Kassanis, 1963). Induced resistance, therefore, is not useful in the field against the economically important systemic virus diseases.

Systemic infections, however, do induce both PR proteins and resistance to localized virus infections (Kassanis *et al.*, 1974), and some of the earlier work on cross-protection has been confused by this phenomenon. Future work on IVR and PR proteins may identify the genes responsible for recognizing and then localizing virus infection. These genes might then be transferred to crop plants.

Inoculants to give plants desirable characteristics, other than resistance to pathogens

As well as deleterious symptoms, virus infection can give desirable changes in growth habit, colour and shape of leaf, flower and fruit. Some ornamental plants owe their attractive foliage to virus infection; for example, abutilon mosaic virus induces a mosaic in *Abutilon* leaves. Tulips (*Tulipa* spp.) with flower breaking were prized possessions before it was established that the colour variations were due to infection with tulip breaking virus. However, selection for breeding stock has usually eliminated plants infected with virus even if the virus gives no apparent symptoms.

Molecular biology has provided the means by which genes may be identified and moved from plant to plant, as long as effective gene vectors are available. Gene transfer by direct insertion of the DNA into the cell by *Agrobacterium* and by plant viruses are currently used. As gene vectors, plant viruses have several advantages. They are easy to inoculate and within a few days spread throughout the plant which remains infected for life. Other methods of gene transfer involve isolating small pieces of tissue or single cells, transforming them and then regenerating from them a complete plant. Many viruses multiply and produce their virus-coded proteins very efficiently. All crop plants are susceptible to some viruses, whereas *Agrobacterium* (Beringer *et al.*, Chapter 8) has a limited host range. A virus gene vector can be selected on its ability to be transferred through seed or pollen.

The main disadvantages of viruses as gene vectors are however equally impressive. Most viruses cause symptoms. Viruses are infectious and many can be easily spread by natural virus vectors, especially invertebrates. Synergistic reactions can occur with other viruses. Most plant viruses have an RNA genome, and the caulimoviruses, one of the two groups of DNA plant viruses, replicate via an RNA intermediate (Hull and Covey, 1983). RNA replication is prone to error and Van Vloten-Doting *et al.* (1985) have suggested that unless there is some selection pressure in favour of the introduced gene it will quickly suffer mutation and be lost. Siegel (1985), however, has described examples were virus-encoded genes, apparently under little selection pressure, have remained unchanged.

Brisson *et al.* (1984) have successfully introduced the gene for methotrexate resistance into cauliflower mosaic virus and found it to be expressed in virus-infected plants. The main use for plant viruses as gene vectors will probably be the fast systemic amplification of a particular coding sequence. Thus plants will be inoculated with a genetically engineered virus for rapid and large-scale production of a desirable product that can then be purified from the infected plants.

Conclusions

Currently the main use of plant viruses as inoculants is to confer resistance against other viruses. There has been little practical application of virus inoculants because of their limited effectiveness, the success in controlling virus diseases by other methods and the dangers of dealing with an infectious pathogen that could be very damaging to other crops. Future prospects, however, are very exciting, particularly with the use of 'partial' inoculants where limited sections of the virus genome are expressed to make the plant resistant to infection. It seems likely that in the future most crop plants will incorporate protection from this source.

The use of viruses as gene vectors is less promising, unless the problems of stability and accidental spread from the target crop can be overcome.

References

Ahl,P. and Gianinazzi,S. (1982) b-Protein as a constitutive component in highly (TMV) resistant interspecific hybrids of *Nicotiana glutinosa* × *Nicotiana debneyi*. *Plant Science Letters*, **26**, 173–181.

Antoniw,J.F., Ritter,C.E., Pierpoint,W.S. and Van Loon,L.C. (1980) Comparison of three pathogenesis-related proteins from plants of two cultivars of tobacco infected with TMV. *Journal of General Virology*, **47**, 79–87.

Antoniw,J.F. and White,R.F. (1986) Changes with time in the distribution of virus and PR protein around single local lesions of TMV-infected tobacco. *Plant Molecular Biology*, **6**, 145–149.

Bar-Jospeh,M. (1978) Cross protection incompleteness: a possible cause for natural spread of citrus tristeza virus after a prolonged lag period in Israel. *Phytopathology*, **68**, 1110–1111.

Barker,H. and Harrison,B.D. (1978) Double infection, interference and superinfection in protoplasts exposed to two strains of raspberry ringspot virus. *Journal of General Virology*, **40**, 647–658.

Baulcombe,D.C., Hamilton,W.D.O., Mayo,M.A. and Harrison,B.D. (1987) Resistance to viral disease through expression of viral genetic material from the plant genome. In *Plant Resistance to Viruses*, Ciba Foundation Symposium 133. Wiley, Chichester, pp. 170–184.

Bol,J.F. (1987) In discussion of Genetic engineering for protection. In *Plant Resistance to Viruses*, Ciba Foundation Symposium 133. Wiley, Chichester, p. 164.

Brisson,N., Pastzkowski,J., Penswick,R., Gronnenborn,B., Potrykus,I. and Hohn,T. (1984) Expression of a bacterial gene in plants by using a viral vector. *Nature*, **310**, 511–514.

Broadbent,L. (1976) Epidemiology and control of tomato mosaic virus. *Annual Review of Phytopathology*, **14**, 75–96.

Channon,A.G., Chettins,N.J., Hitchon,G.M. and Barker,J. (1978) The effect of inoculation with an attenuated mutant strain of tobacco mosaic virus on the growth and yield of early glasshouse tomato crops. *Annals of Applied Biology*, **88**, 121–130.

Costa,A.S. and Muller,G.W. (1980) Tristeza control by cross protection: A U.S.–Brazilian cooperative success. *Plant Disease*, **64**, 538–541.

Dodds,J.A., Lee,S.Q. and Tiffany,M. (1985) Cross protection between strains of cucumber mosaic virus: Effect of host and type of inoculum on accumulation of virions and double-stranded RNA of the challenging strain. *Virology*, **144**, 301–309.

Fletcher,J.T. and Rowe,J.M. (1975) Observations and experiments on the use of an avirulent mutant strain of tobacco mosaic virus as a means of controlling tomato mosaic. *Annals of Applied Biology*, **81**, 171–179.

Fritig,B., Kauffmann,S., Dumas,B., Geoffroy,P., Kopp,M. and Legrand,M. (1987) Mechanism of the hypersensitivity reaction of plants. In *Plant Resistance to Viruses*. Ciba Foundation Symposium 133. Wiley, Chicester, pp. 92–108.

Gianinazzi,S. (1983) Genetic and molecular aspects of resistance induced by infections or chemicals. In Nestor,E.W. and Kosuge,T. (eds), *Plant—Microbe Interactions: Molecular and Genetic Perspectives.* Macmillan, New York, pp. 321—342.

Harrison,B.D., Mayo,M.A. and Baulcombe,D.C. (1987) Virus resistance in transgenic plants that express cucumber mosaic virus satellite RNA. *Nature,* **328,** 799—802.

Hogue,R. and Asselin,A. (1987) Detection of 10 additional pathogenesis-related (b) proteins in intercellular fluid extracts from stressed 'Xanthi-nc' tobacco leaf tissue. *Canadian Journal of Botany,* **65,** 476—481.

Hull,R. and Covey,S.N. (1983) Does cauliflower mosaic virus replicate by reverse transcription? *Trends in Biochemical Sciences,* **8,** 119—121.

Kaper,J.M. and Tousignant,M.E. (1984) Viral satellites: parasitic nucleic acids capable of modulating disease expression. *Endeavour, New Series,* Vol. 8, No. 4.

Kassanis,B. (1963) Interactions of viruses in plants. *Advances in Virus Research,* **10,** 219—255.

Kassanis,B., Gianinazzi,S. and White,R.F. (1974) A possible explanation of the resistance of virus infected tobacco plants to second infection. *Journal of General Virology,* **23,** 11—16.

Kim,S.K. and Wold,B.J. (1985) Stable reduction of thymidine kinase activity in cells expressing high levels of anti-sense RNA. *Cell,* **42,** 129—138.

Loebenstein,G. (1987) In discussion of Resistance and interferon. In *Plant Resistance to Viruses,* Ciba Foundation Symposium 133. Wiley, Chichester, pp. 116—117.

McKinney,H.H. (1929) Mosaic diseases in the Canary Islands, West Africa and Gibraltar. *Journal of Agricultural Research,* **39,** 557—578.

Nelson,R.S., Powell Abel,P. and Beachy,R.N. (1987) Lesions and virus accumulation in inoculated transgenic tobacco plants expressing the coat protein gene of tobacco mosaic virus. *Virology,* **158,** 126—132.

Otsuki,Y. and Takebe,I. (1976) Double infection of isolated tobacco leaf protoplasts by two strains of tobacco mosaic. In Tomiyama,K., Daly,J.M., Uritanai,I., Oku,H. and Ouchi,S. (eds), *Biochemistry and Cytology of Plant Parasite Interaction.* Elsevier, Amsterdam, pp. 213—222.

Parent,J.G. and Asselin,A. (1984) Detection of pathogenesis-related (PR or b) and other proteins in the intercellular fluid of hypersensitive plants infected with tobacco mosaic virus. *Canadian Journal of Botany,* **62,** 564—569.

Pines,O. and Inouye,M. (1986) Anti-sense RNA regulation in prokaryotes. *Trends in Genetics,* **2,** 284—287.

Po Tien, Xiuhua Zhang, Bingsheng Qiu, Bingyi Qin and Gusui Wu (1987) Satellite RNA for the control of plant disease caused by cucumber mosaic virus. *Annals of Applied Biology,* **111,** 143—152.

Ponz,F. and Bruening,G. (1986) Mechanisms of resistance to plant viruses. *Annual Review of Phytopathology,* **24,** 355—381.

Powell Abel,P., Nelson,R.S., De,B., Hoffman,N., Rogers,S.G., Fraley,R.T. and Beachy,R.N. (1986) Delay of disease development in transgenic plants that express the tobacco mosaic virus coat protein gene. *Science,* **232,** 738—743.

Rast,A.T.B. (1972) MII-16, an artificial symptomless mutant of tobacco mosaic virus for seedling inoculation of tomato crops. *Netherlands Journal of Plant Pathology,* **78,** 110—112.

Ross,A.F. (1961a) Localized acquired resistance to plant virus infection in hypersensitive hosts. *Virology,* **14,** 329—339.

Ross,A.F. (1961b) Systemic acquired resistance induced by localized virus-infections in plants. *Virology,* **14,** 340—358.

Ross,A.F. (1966) Systemic effects of local lesion formation. In Beemster,A.B.R. and Dijkstra,J. (eds), *Viruses of Plants.* North Holland Publications, Amsterdam, pp. 127—150.

Russell,G.E. (1970) Interactions between diseases of sugar beet leaves. *National Agricultural Advisory Service Quarterly Review,* **87,** 132—138.

Shalla,T.A. and Peterson,L.J. (1978) Studies on the mechanism of viral cross protection. *Phytopathology,* **68,** 1681—1683.

Sherwood,J.L. and Fulton,R.W. (1982) The specific involvement of coat protein in tobacco mosaic virus cross protection. *Virology,* **119,** 150—158.

Siegel,A. (1985) Plant-virus-based vectors for gene transfer may be of considerable use despite a presumed high error frequency during RNA synthesis. *Plant Molecular Biology,* **4,** 327—329.

Tumer,N.E., O'Connell,K.M., Nelson,R.S., Sanders,P.R., Beachy,R.N., Fraley,R.T. and Shah,D.M. (1987) Expression of alfalfa mosaic virus coat protein gene confers cross-protection in transgenic tobacco and tomato plants. *The European Molecular Biology Organization Journal,* **6,** 1181—1188.

Van Dun,C.M.P., Bol,J.F. and Van Vloten-Doting,L. (1987) Expression of alfalfa mosaic virus and tobacco rattle virus coat protein genes in transgenic tobacco plants. *Virology,* **159,** 299—305.

Van Loon,L.C. (1985) Pathogenesis-related proteins. *Plant Molecular Biology,* **4,** 111—116.

Van Vloten-Doting,L., Bol,J. and Cornelisson,B. (1985) Plant-virus-based vectors for gene transfer will be of limited use because of the high error frequency during vital RNA synthesis. *Plant Molecular Biology,* **4**, 323−326.

White,R.F., Dumas,E., Shaw,P. and Antoniw,J.F. (1986) The chemical induction of PR (b) proteins and resistance to TMV infection in tobacco. *Antiviral Research,* **6**, 177−185.

White,R.F., Rybicki,E.P., Von Wechmar,M.B., Dekker,J.L. and Antoniw,J.F. (1987) Detection of PR1-type proteins in *Amaranthaceae, Chenopodiaceae, Graminae* and *Solanaceae* by immunoelectroblotting. *Journal of General Virology,* **68**, 2043−2048.

Yarwood,C.E. (1960) Localized acquired resistance to tobacco mosaic virus. *Phytopathology,* **50**, 741−744.

CHAPTER 7

Use of blue-green algae and *Azolla* in rice culture

B.A.WHITTON and P.A.ROGER[1]

Department of Biological Sciences, University of Durham, Durham DH1 3LE, UK and [1]International Rice Research Institute, P.O. Box 933, Manila, Philippines

Introduction

Blue-green algae (cyanobacteria) are distributed world-wide and contribute to the fertility of many agricultural ecosytems, either as free-living organisms or in symbiotic association with the water-fern *Azolla* (Fay, 1983). The nitrogen-fixing ability of many species is the principal, but by no means the only, reason for this increased fertility. The particular importance of these organisms in rice culture was made clear in the review by Roger and Kulasooriya (1980). This included many reports of the manipulation of rice field ecosystems to maximize blue-green algal nitrogen fixation, especially by the deliberate addition of dried inocula. However, most of these reports lacked detailed documentation of methods and results, and a most recent review (Roger, 1989) takes a cautious view when interpreting the significance of earlier research. The literature on the use of *Azolla* is, however, much more detailed (Lumpkin and Plucknett, 1982; Shi and Hall, 1988).

This chapter examines the possibilities for deliberate modifications of blue-green algal populations in rice fields, with the ultimate aim of increasing rice yield. Topics dealt with in detail by Roger (1989) are treated here only briefly, especially those concerning *Azolla*.

Free-living blue-green algae

Occurrence and agronomic significance

The abundance of blue-green algae in rice fields has been reported in numerous papers since Fritsch's accounts (Fritsch, 1907a,b). Culture studies were introduced by Bannerji (1935) and the importance of blue-green algal nitrogen fixation in helping to maintain fertility of the rice fields was first recognized by De (1939). Many rice fields show visually obvious growths of blue-green algae, although eukaryotic green algae may be more abundant where high quantities of nitrogenous fertilizer have been added. Reports from many countries indicate that the blue-green algal flora is often rich in species (Gupta, 1966; Ali *et al.*, 1978; Saha and Mandal, 1979; Anagnostidis *et al.*, 1981; Al-Mousawi and Whitton, 1983; Kulasooriya and de Silva, 1981). Typically, about half the blue-green algal genera represented are heterocystous (*Anabaena, Aulosira,*

89

Calothrix, Cylindrospermum, Fischerella, Gloeotrichia, Nostoc, Scytonema, Tolypothrix, Wollea) and thus nitrogen-fixers. A quantitative study (Roger *et al.*, 1987) of 102 samples of rice soils from the Philippines, India, Malaysia and Portugal showed that heterocystous blue-green algae occurred at densities ranging from 10^2 to 8×10^6 colony-forming units (CFU) per cm^2. The abundance of heterocystous forms shows a positive correlation with pH and available phosphorus content of soils. Some communities dominated by non-heterocystous forms also fix nitrogen, though laboratory studies show that associated bacteria are sometimes responsible for the fixation (R.Islam and B.A.Whitton, unpublished).

In contrast to most subsequent studies, Watanabe *et al.* (1951) found only a low number (13) of species of nitrogen-fixing blue-green algae in 643 rice field samples from various parts of Asia. In a subsequent study, Watanabe (1959) reported that there was a complete absence of blue-green algae in many rice field soils, such as the Kanto loams of Japan, which have pH values of 5−6. However, more recent studies have shown the consistent presence of nitrogen-fixing blue-green algae, frequently at high densities, in soils under rice cultivation (Roger *et al.*, 1987). It remains uncertain whether the soils tested by Watanabe (1959) really were atypical, but the culture medium given by Watanabe *et al.* (1951) appears to be unsuitable for the isolation of nitrogen-fixing blue-green algae.

The best known effect of blue-green algal growth on rice increased nitrogen availability resulting from nitrogen fixation but other effects have been reported. They may prevent the growth of weeds (Subrahmanyan *et al.*, 1965) and they add to the soil organic content, aiding particle aggregation (Roychoudhury *et al.*, 1980). Increased availability of phosphorus to rice found by Arora (1969) was explained by the excretion of organic acids by blue-green algae. The presence of blue-green algae in the immediate vicinity of rice seeds can decrease sulphide injury; this can be achieved either by pre-soaking the seeds in blue-green algal cultures (Jacq and Roger, 1977) or by field inoculation (Aiyer *et al.*, 1971). There have been many claims that soaking seeds or joint growth of seedlings with blue-green algal cultures can benefit rice plants by producing plant growth regulators. Roger and Kulasooriya (1980) listed many physiological and morphological responses of rice plants which have been interpreted by the original authors in this way. However, when 133 cyanobacterial isolates from sites in Africa (not all rice fields) were tested (Pedurand and Reynaud, 1987) for their effects on rice germination and growth, 70% had a negative effect on germination and only 21% a stimulatory effect; many *Nostoc* strains had a negative effect. They concluded that presoaking rice seeds in a blue-green algal culture should be done with caution or avoided altogether. No study has been made in which a plant growth regulator has been isolated and characterized from blue-green algal material (Metting and Pyne, 1986).

The extent to which the blue-green algae may contribute to the nitrogen requirements of the rice crop is determined by a number of factors, the most obvious of which are the standing crop, rate of nitrogen fixation per unit area, turnover of the nitrogen fixed and the extent to which any nitrogen released becomes available to the rice plant. The relevant literature, summarized by Roger (1989), indicates that relatively little is known about nitrogen turnover or the extent to which this nitrogen becomes available to the plant. Standing crops of nitrogen-fixing blue-green algae range from a few kg to 0.5 t ha^{-1} dry weight (Roger *et al.*, 1987) and the various lines of evidence indicate a potential of approximately 30 kg N ha^{-1} per crop. Nitrogen released in the later part of the growth period of the rice crop may be too late to influence grain yield, though it may be important for the subsequent agricultural crop. Uptake of ^{15}N by rice from blue-green algae has been the subject of several studies. For instance, in pot and field trials by Tirol *et al.* (1982), 23−28% of the nitrogen fixed reached the first crop; ^{15}N from blue-green algae in a deepwater rice plot has also been shown to reach the rice plant (Watanabe and Ventura, 1982).

Methods of increasing blue-green algal biomass

The fact that different rice fields in the same region and at the same time may have markedly different blue-green algal standing crops suggests that different farming practices may influence their development. For instance, deepwater rice fields at two sites in Bangladesh shortly before the arrival of floodwater showed a marked contrast (B.A.Whitton, unpublished). Bunds (small embankments) forming the margins of fields at one site were continuous and therefore held rainwater, whereas those at another site were broken, permitting drainage from the fields. Fields at the former site had an abundant algal cover, whereas visually obvious growths were rare at the latter site. Such observations raise the possibility of deliberate manipulation of the ecosystem to favour blue-green algae by liming, phosphorus application, surface application of straw and grazer control (Watanabe *et al.*, 1981; Grant *et al.*, 1983). The addition of nitrogen fertilizer has been shown in a number of studies to decrease blue-green algal growth (Roger and Kulasooriya, 1980) or influence species composition, but the results appear somewhat erratic. Possible reasons for this are the frequent simultaneous addition of phosphate and rapid mobilization of the nitrogen in the soil. Deep placement of nitrogen fertilizer reduces its inhibitory effect on blue-green algal nitrogen fixation (Roger *et al.*, 1980).

Floating gelatinous colonies of *Nostoc* are added to some fields in China in much the same way as *Azolla* (see below), with populations allowed to develop in ponds and then released into paddy fields when the rice is planted (T.A.Lumpkin, pers. comm.). This system appears to be only local. Pantastico and Gonzales (1986) reported experimental studies with similar *Nostoc* in the Philippines which led to an increase in grain yield up to 22%. Growth of the blue-green alga was influenced markedly by grazers, but use of pot trials with a mixed system involving *Nostoc*, rice and *Tilapia* overcame this problem (Martinez *et al.*, 1978).

The effects of rice yield of soil inoculation by blue-green algae were first reported by Watanabe *et al.* (1951), with a 25% increase in yield after inoculation of poorly drained paddy with *Tolypothrix tenuis*. Several authors reported increases well over 200% from pot trials in India (Singh, 1961; Sundara Rao *et al.*, 1963). Almost all subsequent studies have indicated much lower increases in the field than pot trials, even where comparative studies have been made by the same researcher (Huang Chi-ying, 1978). Studies on the use of inocula for rice soils have been discontinued in Japan, but subsequently there have been many reports from India and a limited number from other countries. Inocula have mostly been derived from laboratory-grown strains, following the early studies with *T.tenuis* in Japan (Watanabe, 1962, 1973).

The interest generated in India led in 1977 to the All-India Coordinated Project on

Algae, which involves the production and distribution of inocula. Books based on the results give details of practical methods (Venkataraman, 1972, 1981). Inocula are derived from a mix of strains isolated originally from rice-fields and grown in shallow trays with soil, phosphate and insecticide. If necessary, lime is added to adjust soil pH to 7.0−7.5. The blue-green algal mats which develop are allowed to dry and the dried flakes are stored in bags for use at 10 kg ha^{-1} in farmers' fields. Algalization, the term widely used for the addition of such inocula, has received considerable publicity. Some reviews (Agarwal, 1979) have accepted the success of the method in raising grain yield as a well-established fact. Many studies have reported increased grain yield, grain nitrogen content or straw nitrogen content (Venkataraman, 1981; Singh and Singh, 1987), with the effects of blue-green algae being equivalent to the addition of 20−30 kg ha^{-1} nitrogen provided phosphorus fertilizer is added (Sharma and Gupta, 1983). Reports on field experiments available to Roger and Kulasooriya (1980) showed that on average, algal inoculation, where effective, causes about 14% increase in grain yield, corresponding to about 450 kg grains ha^{-1} per crop. However, Roger (1989) concludes that the effects of inoculation of rice-fields by free-living blue-green algae seem often to be erratic and limited. It is difficult to find a clear-cut example which shows the increase to be statistically significant. Firm data also appear to be lacking to support the statement from Venkataraman (1981) and quoted by other authors (Metting, 1988): 'A conservative estimate suggests that about two million hectares under rice are currently covered with algal biofertilizer technology'. Roger *et al.* (1985) reported that algalization was adopted in only two states of India and there the inoculated fields comprise only a few percent of the total area under rice.

A deliberate increase in blue-green algal population density is likely to be much more important where there are marked seasonal changes in use of land, such as when the ground is ploughed many times before planting a winter crop other than rice. Under such circumstances the natural blue-green algal population density may be low at the beginning of the subsequent rice season, leading to a lag of several weeks before it can make a significant contribution to nitrogen fixation. Under some circumstances, inoculation may be more effective if carried out after the rice has been planted. For instance, a multivariate analysis of data from West African sites (Reynaud, 1987) showed that the best time to inoculate is at the beginning of tillering. There are other ways in which algalization may help. For instance, added inocula may have an advantage over the *in-situ* algal population at the start of the rice season because of the use of phosphorus fertilizer during the production of the inoculum. The effect is, however, only likely to be more important if the inoculum is derived from local populations (Bisoyi and Singh, 1989). Cells in the inoculum are probably phosphorus-rich and can therefore divide rapidly and perhaps increase in density by an order of magnitude independently of soil phosphorus. This is likely to be more important in the case of those species which survive desiccation as whole filaments, as akinetes (spores) are apparently not characterized by a high phosphorus content (Whitton, 1987).

Most algalization trials have been carried out using inocula developed from mixes of laboratory isolates. The data appear to be lacking to support the claim by Agarwal (1979) that the (introduced) blue-green algae can establish themselves almost permanently if inoculation is done repeatedly for 3−4 cropping seasons. The only quantitative study to establish the fate of strains subsequent to inoculation is that of Reddy and Roger (1988). In this study the fate of five laboratory-grown heterocystous strains representing 75% of the inoculum was studied for 1 month in 1 m^2 plots of five different soils. During the month following inoculation, the inoculated strains multiplied to some extent in all soils, but rarely dominated the indigenous blue-green algae and did so only when the growth of indigenous nitrogen-fixing species was poor or after population declines of indigenous species. The soils were dried at the end of the period and then resubmerged, together with neem (*Azadirachta indica*), to control grazers. Two of the inoculated strains did not reappear, but one (*Aulosira fertilissima*) developed an agronomically significant population on two soils. In field situations with a rich natural flora, it seems likely that indigenous strains will usually rapidly outgrow populations derived from the original laboratory isolates. Where farmers increase their inocula in the shallow trays, it is probable that strains present in the added local soil may outgrow the original laboratory isolates even before inocula are added to the fields.

Efforts have been made to obtain strains with especially high nitrogen-fixing ability, so that these can be incorporated into the inocula. The approaches used have included screening of a range of strains obtained from enrichment cultures and attempts to obtain mutants in strains already in culture. The former approach has provided strains which are fast-growing in the laboratory (Antarikanonda and Lorenzen, 1982), but there is as yet no evidence that such strains have a particular advantage in the field. Although there are few quantitative data, many rice field blue-green algae probably often double every 1–3 days. It seems unlikely, therefore, that an introduced strain will survive long in competition with the natural flora, unless there is a simultaneous change in the environment which gives it a competitive advantage. It may prove useful to introduce fast-growing strains, if there is a sharp change in fertilizer practice, such as the use of high phosphorus, but no added source of nitrogen at a site which was not previously fertilized.

A further use for selected strains might be the addition of inocula resistant to pesticides. There are marked differences in the relative sensitivity of blue-green algae and weeds to widely used herbicides; the former are sometimes relatively insensitive. Overall, however, it is clear that the growth and activities of blue-green algae are affected adversely by some commonly used pesticides (Singh and Singh, 1983; Padhy, 1985b). A nitrogen-fixing strain of *Gloeocapsa* isolated from a rice-field was reported by Singh et al. (1986) to be highly resistant to the herbicides Machete and Basalin, whereas *Nostoc muscorum* from another source was quite sensitive. Repeated laboratory culture with increasing levels of pesticide led to increased resistance of three nitrogen-fixing strains to four fungicide and insecticides (Sharma and Gaur, 1981). Artificially induced mutants resistant to Blitox have been obtained for two *Nostoc* strains, and other blue-green algal mutants have obtained for resistance to Carbaryl, Zineb and Mancozeb (Padhy, 1985a). Most studies on pesticide tolerance in blue-green algae have been laboratory-based and without regard to the source of the strains tested. It would be useful to have more data based directly on field observations and assays on strains of particular species taken from sites with a known pesticide history. This would indicate whether or not genetic tolerance to particular pesticides is acquired by blue-green algae easily under field conditions.

B.A. Whitton and P.A. Roger

Azolla

Occurrence and agronomic aspects

Abundant growths of *Azolla* not only make a useful addition of combined nitrogen to the ecosystem but can also provide a 'green manure'. In contrast to free-living blue-green algae, however, *Azolla* usually needs to be inoculated and cultivated to develop a significant biomass in rice fields (Roger, 1989). According to Lumpkin and Plucknett (1980, 1982), there is a long history of *Azolla* cultivation in China and Vietnam, records for the latter going back to the 11th century. There was a research thrust in both countries in the 1960s, combined with a considerable expansion in cultivation, and this interest has spread to a number of other countries since the 1970s. The *Azolla* can be intercropped with other plants besides rice, such as *Sesbania* and *Colocasia* (Kannaiyan, 1987), but most interest has been focused on rice. The recent introductions to other countries have usually commenced with trials at research institutes or similar organizations. In the South Cotabato region of the Philippines, *Azolla* was used on 84 000 ha in 1985 (Mabbayad, 1987) and in some other countries, such as India (Kannaiyan, 1987; P.K.Singh, pers. comm.), mixed *Azolla*−rice cultivation also appears to be sufficiently well established in a few regions to exist without the support of agricultural extension workers, but in others such as Brazil (Fiore and Gutbrod, 1987), it is probably still almost entirely dependent on such support.

How important is mixed *Azolla*−rice cultivation on a global scale? Data in the literature for China, by far the most important country for *Azolla*, include markedly different values, but Lumpkin and Plucknett (1982) gave an estimate of 2% of the 34 million ha of rice. The use of *Azolla* in China as a green manure is decreasing (Liu Chung-chu, pers. comm. to P.A.Roger). *Azolla* is also widely used for other purposes, such as feeding pigs, ducks and cattle, in China and some other countries in the region (Moore, 1969; Edwards, 1980) and there is increasing interest in its potential as food for fish (Antoine *et al.*, 1986), with an integrated rice−*Azolla*−fish system developed in China (Liu Chung-chu, 1987). However, it seems probable that *Azolla* is currently used on less than 2% of the global 150 million ha under rice cultivation (Roger, 1989).

The system of utilizing the *Azolla* differs slightly from place to place, but whatever is done is labour-intensive (Lumpkin and Plucknett, 1982). Typically the plant is inoculated into flooded fields before the rice is transplanted, but this may also be done at the same time as transplanting. After about 2 weeks the water may be partially drained away and the *Azolla* heeled or ploughed into the soil. In some cases part of the *Azolla* crop is removed and placed in piles to develop into organic fertilizer, which is subsequently incorporated into the soil. This process may be repeated, but eventually the field is left undisturbed, the rice planted, if not already done, and the *Azolla* grows intermingled with the rice and any weeds. Inocula are developed in ponds, which typically received phosphorus fertilizer, sometimes other nutrients, and usually an insecticide. More fertilizer may be added to the intercrop *Azolla*.

Quantitative studies on *Azolla* biomass and its contribution to the nitrogen economy of rice-fields have been reviewed by Roger and Watanabe (1986) and Watanabe (1987). The reported maximum standing crop of *Azolla* ranged from 0.8 to 5.2 t ha^{-1} and averaged 2.1 t ha^{-1}. International field trials conducted for four years at 37 sites in 10 countries (Watanabe, 1987) showed that incorporation of one crop of *Azolla* grown before or after transplanting is equivalent to a split application of 30 kg fertilizer N; incorporation of two *Azolla* crops grown before or after transplanting is equivalent to split application of 60 kg nitrogen.

Environmental factors and use of Azolla

Successful growth of *Azolla* is more demanding than that for a blue-green algal cover. The environmental factors which restrict the area over which *Azolla* is used are reviewed by Lumpkin (1987). The most obvious factor is water, because *Azolla* plants can survive only a few days on paddy soil that dries during intermittent rains (Lumpkin, 1987). This is in marked contrast to rice field blue-green algae, almost all species of which are tolerant of drying, often without the need for akinete formation (authors' unpublished data). *Azolla* can form sporocarps which survive drying, but the factors which lead to sporulation are not well understood, although it is now possible to produce sporocarps routinely with *A. filiculoides* (Shuying, 1987) and the 'red duckweed' *Azolla* sp. (Xiao Qing-yuan, 1987). Conditions favouring germination of sporocarps are much better known (Xiao Qing-yuan *et al.*, 1987) than are those which lead to their formation.

Azolla has a relatively high requirement for light. When it is grown as an intercrop with rice, its growth will start to be influenced by the rice leaf canopy 2−3 weeks after transplanting and it will stop growth 45 days after transplanting in most *Azolla* species (Lumpkin, 1987). Temperature is an especially important limiting factor. The optimum temperature for most species is within the range 20−35°C (Lumpkin, 1987), though some *Azolla* strains can grow at temperatures of 40°C or more (Watanabe and Berja, 1983). The poor growth of *Azolla* at higher temperatures is at least in part due to increased effects of grazing, parasites and competition with free-living algae. *Azolla* appears to be most successful in the subtropics, with the optimum temperature for most species in the 20 30°C range. Lumpkin (1987) comments that because high temperatures are not a direct limitation, *Azolla* has an excellent potential for successful cultivation in irrigated deserts where humidity is relatively low and alternate host plants for insects are limited; he mentions the northern border of Senegal as a region where it already does well. This suggests its possible use in many countries for which there are at present few or no reports, such as parts of North Africa and the Middle East.

Phosphorus is typically the major limiting element in the field (Roger and Watanabe, 1986; Roger, 1989). As the plant can accumulate about 10 times the amount of phosphorus required to support its normal nitrogen concentration (Lumpkin, 1987), the use of phosphorus fertilizer in ponds used for developing inocula permits the plant to increase in biomass when transferred to paddy fields, even if water in the field is low in phosphorus (Watanabe *et al.*, 1988).

Possible new symbiotic associations

The most direct method to enhance blue-green algal nitrogen fixation in rice fields would be to produce a new symbiotic association combining rice and a suitable nitrogen-fixing strain. Is this approach plausible? Although the *Azolla*−*Anabaena azollae* association is at present the only symbiotic association important in rice culture, examples of symbiotic associations between blue-green algae and other organisms are widespread in nature. There are records from most phyla, both plant and animal, and more and more examples are being found. The blue-green alga may be intra- or extracellular and in most cases is a nitrogen-fixer. This has led to attempts over the past 20 years to develop new symbiotic associations combining a blue-green alga with a crop plant. Several research groups have succeeded in incorporating a unicellular blue-green alga into higher plant protoplasts, though apparently not yet a nitrogen-fixing strain and not yet in a long-term stable relationship (Gamborg and Bottino, 1981). An attempt to produce tobacco protoplasts incorporating the filamentous nitrogen-fixer, *Anabaena variabilis*, was relatively unsuccessful (Meeks et al., 1978); the protoplasts disintegrated in five days, although some blue-green algal filaments remained intact.

Systems for mixed culture of a blue-green alga and higher plant tissue have been developed by a group at Moscow State University (Gusev et al., 1984), though rice has apparently not been tested. Mixed cultures of tobacco callus and *A. variabilis* were shown (Gusev et al., 1986) to be capable of nitrogenase activity (acetylene reduction assay), and regenerated tobacco plants have been obtained with *A. variabilis* in the intercellular spaces of the primary cortex (Pivovarova et al., 1986). It is not essential for the blue-green alga to act as an autotroph in such relationships, since *Chlorogloea* (*Chlorogloeopsis*) *fritschii* is able to grow in mixed culture in the dark (Gusev et al., 1980). In the naturally occurring symbiotic associations involving *Gunnera* and *Cycas*, the nitrogen-fixing blue-green alga exists entirely or largely in the dark, presumably obtaining all its fixed carbon from the host plant. This indicates an important advantage that a symbiotic blue-green algal−rice system might have over the use of free-living nitrogen-fixers. Fixation would not be limited by the lack of light, as the leaf canopy increased during the growth of the crop but would take photosynthate from the crop.

When considered together, the evidence summarized above suggests that it is quite feasible that some form of artificial symbiotic association will one day be created between a rice plant and a nitrogen-fixing blue-green alga and that the alga need not necessarily be located in the light. Whether or not it proves to have any practical relevance is another matter. In view of the importance of input of combined nitrogen to rice-fields and the fact that blue-green algae are often abundant in this ecosystem, there would seem to have been plenty of opportunity for symbiotic associations to have evolved naturally. The abundance of blue-green algae on aquatic roots of wild deepwater rice in well-illuminated situations (authors' observations) indicate that the close association of rice plants with blue-green algae occurred long before the plant was cultivated. However, the only report of endophytic algal growth is for *Nostoc* and *Calothrix* inside senescent leaf sheaths of cultivated deepwater rice plants (Kulasooriya et al., 1980; Whitton et al., 1989). Even if all the nitrogen fixed by these organisms passed to the rice plant, it would probably supply less than 0.05% of the plant's requirements (B.A.Whitton, unpublished). This suggests that there has been no marked pressure favouring the evolution of a rice−blue-green algal symbiotic association.

Conclusions

We will summarize by stating what we believe is the situation. It should be possible to influence the amount of blue-green algal nitrogen fixation in many rice fields. The response is likely to be most evident in non-acidic soils not treated with nitrogenous fertilizer with moderate to high phosphorus availability. One of the ways of achieving an increase in the blue-green algal population is by addition of inocula, a method which is certainly sometimes effective. However, in contrast to the many excellent studies on *Azolla*, such as those carried out by the Central Rice Research Institute in India, many of the studies on algalization have been uncritical. Further, they have often appeared to be concerned only with grain yield and to lack any interest in the microbial and ecological processes that might influence this. The extent to which algalization will eventually prove worthwhile will depend on local circumstances. In general, free-living blue-green algae have less potential in terms of nitrogen fixed than *Rhizobium*, legumes or *Azolla* (Roger and Watanabe, 1986), but need less input of time and materials.

The dearth of quantitative studies on the fate of introduced inocula of free-living blue-green algae is unfortunate, not only because it has hindered the evaluation of algalization studies, but also because of the current interest in the fate of added microbial inocula to ecosystems associated with the release of genetically engineered organisms. Blue-green algae are potentially one of the most useful groups for such studies, because it is possible to use strains with easily detectable morphological features such as spore coats.

Although the spread of *Azolla* culture resulting from recent research has been quite slow, we suggest that it will eventually spread throughout the subtropics, except for countries like Australia and the USA where labour costs are high. Expansion is likely to be favoured by the breeding of strains which overcome environmental problems, and the increasing use of the plant for purposes other than green manure, such as animal feed. However, the area incorporating *Azolla* cultivation is unlikely much to exceed 2% of the total rice area, unless the costs of nitrogen fertilizers increase markedly or strains can be obtained which thrive under fully tropical conditions.

It seems quite probable that success will one day be achieved in bringing about a symbiotic association between a rice cultivar and a nitrogen-fixing organism and that the latter may be a blue-green alga. However, as suggested above, the question is whether such an association would have much practical relevance for rice culture, although it might have for other crops. If it was adopted, it might still be worthwhile to encourage the growth of nitrogen-fixing blue-green algae or *Azolla* before the rice leaf canopy thickens.

Acknowledgements

B.A.Whitton thanks the UK Overseas Development Agency for two research contracts concerning deepwater rice in Bangladesh. P.A.Roger carried out research under a scientific agreement between IRRI and ORSTOM (France), and is grateful to Dr I.Watanabe (IRRI) for helpful discussion.

References

Agarwal,A. (1979) Blue-green algae to fertilize Indian rice paddies. *Nature*, **279**, 181.
Aiyer,R.S., Aboobaker,V.O. and Subramoney,N. (1971) The role of blue-green algae in suppressing sulphide injury to rice crop in submerged soils. *Madras Agriculture*, **58**, 405–407.
Ali,S., Rajoka,M.I. and Sandhu,G.R. (1978) Blue-green algae of different rice growing soil series of the Punjab. *Pakistan Journal of Botany*, **10**, 197–207.
Al-Mousawi,A.H.A. and Whitton,B.A. (1983) Influence of environmental factors on algae in rice-field soil from the Iraqi marshes. *Arab Journal of Gulf Research*, **1**, 237–253.

Anagnostidis,K., Economou-Amilli,A. and Tsangridis,A. (1981) Taxonomic and floristic studies on algae from rice-fields of Kalochorion-Thessaloniki, Greece. *Nova Hedwigia*, **34**, 1–189.

Antarikanonda,P. and Lorenzen,H. (1982) N_2-fixing blue-green algae (Cyanobacteria) of high efficiency from paddy soils of Bangkok, Thailand: Characterization of species and N_2-fixing capacity in the laboratory. *Archiv für Hydrobiologie, Supplemente*, **63**, 53–70.

Antoine,T., Carraro,S., Micha,J.-C. and van Hove,C. (1986) Comparative appetency for *Azolla* of *Cichlasoma* and *Oreochromis* (*Tilapia*). *Aquaculture*, **53**, 95–99.

Arora,S.K. (1969) The role of algae on the availability of phosphorus in paddy fields. *Riso*, **18**, 135–138.

Bannerji,J.C. (1935) On algae found in soil samples from an alluvial paddy field of Faridpur, Bengal. *Science and Culture*, **1**, 298–299.

Bisoyi,R.N. and Singh,P.K. (1989) Effect of phosphorus fertilization on blue-green algal inoculum production and nitrogen yield under field conditions. *Biology and Fertility of Soils*, **5**, 338–343.

De,P.K. (1939) The role of blue-green algae in nitrogen fixation in rice fields. *Proceedings of the Royal Society of London, Series B*, **127**, 121–139.

Edwards,P. (1980) *Food Potential of Aquatic Macrophytes*, International Center for Living Aquatic Resources Management, Manila, Philippines.

Fay,P. (1983) *The Blue-greens*. Edward Arnold, London.

Fiore,M.F. and Gutbrod,K.G. (1987) Use of *Azolla* in Brazil. In *Azolla Utilization. Proceedings of the Workshop on Azolla Use, Fuzhou, Fujian, China, 31 March–5 April 1985*. International Rice Research Institute, PO Box 933, Manila, Philippines, pp. 123–130.

Fritsch,F.E. (1907a) The subaerial and freshwater algal flora of the tropics. *Annals of Botany*, **30**, 235–275.

Fritsch,F.E. (1907b) A general consideration of aerial and fresh water algal flora of Ceylon. *Proceedings of the Royal Society of London, Series B*, **11**, 79–197.

Gamborg,O.L. and Bottino,P.J. (1981) Protoplasts in genetic modifications of plants. In *Advances in Biochemical Engineering*, **19**, (Ed. A.Fiechter). Springer Verlag, Berlin, pp. 239–263.

Grant,I.F., Tirol,A.C., Aziz,T. and Watanabe,I. (1983) Regulation of invertebrate grazers as a means to enhance biomass and nitrogen fixation of cyanophyceae in wetland rice fields. *Soil Science Society of America Journal*, **47**, 669–675.

Gupta,A.B. (1966) Algal flora and its importance in the economy of rice fields. *Hydrobiologia*, **28**, 213–222.

Gusev,M.V., Butenko,R.G. and Korzhenevskaya,T.G. (1984) Cyanobacteria in association with cultivated cells of higher plants. *Soviet Scientific Reviews, Section D*, 1–40.

Gusev,M.V., Butenko,R.G., Korzhenevskaya,T.G., Lobakova,E.D. and Baulina,O.I. (1980) Intercellular symbiosis of suspension ginseng culture cells and cyanobacteria. *European Journal of Cell Biology*, **22**, 503.

Gusev,M.V., Korzhenevskaya,T.G., Pyvovarova,L.V., Baulina,O.I. and Butenko,R.G. (1986) Introduction of a nitrogen-fixing cyanobacterium into tobacco shoot regenerates. *Planta*, **167**, 1–8.

Huang Chi-ying (1978) Effects of nitrogen fixing activity of blue-green algae on the yield of rice plants. *Botanical Bulletin of Academia Sinica*, **19**, 41–52.

Jacq,V. and Roger,P.A. (1977) Diminution des fontes de semis dues à la sulfatoréduction, par un prétraitement des graines de riz avec des cyanophycées. *Cahiers ORSTOM Serie Biologie*, **12**, 101–107.

Kannaiyan,S. (1987) Use of *Azolla* in India. In *Azolla Utilization. Proceedings of the Workshop on Azolla Use, Fuzhou, Fujian, China, 31 March–5 April 1985*, International Rice Research Institute, PO Box 933, Manila, Philippines, pp. 109–118.

Kulasooriya,S.A., Roger,P.A., Barraquio,W.L. and Watanabe,I. (1980) Biological nitrogen fixation by epiphytic microorganisms in rice fields. *IRRI Research Papers*, **47**, 1–10. (International Rice Research Institute, PO Box 933, Manila, Philippines).

Kulasooriya,S.A. and de Silva,R.S.Y. (1981) Multivariate interpretation of the distribution of nitrogen-fixing blue-green algae in rice soils in central Sri Lanka. *Annals of Botany, New Series*, **47**, 31–52.

Liu Chung-Chu (1987) Reevaluation of *Azolla* utilization in agricultural production. In *Azolla Utilization. Proceedings of the Workshop on Azolla Use, Fuzhou, Fujian, China, 31 March–5 April 1985*. International Rice Research Institute, PO Box 933, Manila, Philippines, pp. 67–76.

Lumpkin,T.A. (1987) Environmental requirements for successful *Azolla* growth. In *Azolla Utilization. Proceedings of the Workshop on Azolla Use, Fuzhou, Fujian, China, 31 March–5 April 1985*. International Rice Research Institute, PO Box 933, Manila, Philippines, pp. 89–97.

Lumpkin,T.A. and Plucknett,D.L. (1980) *Azolla*: botany, physiology, and use as a green manure. *Economic Botany*, **34**, 111–153.

Lumpkin,T.A. and Plucknett,D.L. (1982) *Azolla as a Green Manure: Use and Management in Crop Production*. Westview Tropical Agriculture Series, Westview Press, Boulder, CO, USA.

Mabbayad,B.B. (1987) The *Azolla* program of the Philippines. In *Azolla Utilization. Proceedings of the*

Workshop on Azolla Use, Fuzhou, Fujian, China, 31 March – 5 April 1985. International Rice Research Institute, PO Box 933, Manila, Philippines, pp. 101 – 108.
Martinez,M.R., Evangelista,C.L. and Pantastico,J.B. (1978) *Nostoc commune* Vauch, as a potential fertilizer in rice-fish culture: a preliminary study. *The Philippine Journal of Crop Science*, **2**, 251 – 256.
Meeks,J.C., Malmberg,R.L. and Wolk,C.P. (1978) Uptake of auxotrophic cells of a heterocyst-forming cyanobacterium by tobacco protoplasts and the fate of their association. *Planta*, **133**, 55 – 60.
Metting,B. (1988) Micro-algae in agriculture. In Borowitzska,M.A. and Borowitzska,L.J. (eds), *Micro-algal Biotechnology*. Cambridge University Press, Cambridge, pp. 288 – 304.
Metting,B. and Pyne,J.W. (1986) Biologically active compounds from microalgae. *Enzyme and Microbial Technology*, **8**, 385 – 448.
Moore,A.W. (1969) *Azolla*: biology and agronomic significance. *Botanical Reviews*, **35**, 17 – 35.
Padhy,R.N. (1985a) Cyanobacteria employed as fertilizers and waste disposers. *Nature*, **317**, 475 – 476.
Padhy,R.N. (1985b) Cyanobacteria and pesticides. *Residue Reviews*, **95**m, 1 – 44.
Pantastico,J.B. and Gonzales,J.L. (1976) Culture and use of *Nostoc commune* as biofertilizer. *Kalikasan, Philippines Journal of Biology*, **5**, 221 – 234.
Pedurand,P. and Reynaud,P.A. (1987) Do cyanobacteria enhance germination and growth of rice? *Plant and Soil*, **101**, 235 – 240.
Pivovarova,L.V., Korzhenevskaya,T.G., Butenko,R.G. and Gusev,M.V. (1986) Localization of cyanobacteria growing in association with callus culture and with regenerated plants of tobacco. *Akademia Nauk USSR*, **33**, 74 – 81 (in Russian).
Reddy,P.M. and Roger,P.A. (1988) Dynamics of algal populations and acetylene reducing activity in five rice soils inoculated with blue-green algae. *Biology and Fertility of Soils*, **6**, 14 – 21.
Reynaud,P.A. (1987) Ecology of N_2-fixing cyanobacteria in dry tropical habitats of West Africa: a multivariate analysis. *Plant and Soil*, **98**, 203 – 220.
Roger,P.A. (1989) Blue-green algae (cyanobacteria) in agriculture. In Dawson,J.O. and Dart,P. (eds), *Microorganisms that Promote Productivity*. Martinus Nijhoff, Dordrecht.
Roger,P.A., Grant,I.F. and Reddy,P.M. (1985) Blue-green algae in India: a trip report. International Rice Research Institute, Manila.
Roger,P.A. and Kulasooriya,S.A. (1980) *Blue-green Algae and Rice*. International Rice Research Institute, PO Box 933, Manila, Philippines.
Roger,P.A., Kulasooriya,S.A., Tirol,A.C. and Craswell,E.T. (1980) Deep placement: a method of nitrogen fertilizer application compatible with algal nitrogen fixation in wetland rice soils. *Plant and Soil*, **57**, 137 – 142.
Roger,P.A., Santiago-Ardales,S., Reddy,P.M. and Watanabe,I. (1987) The abundance of heterocystous blue-green algae in rice soils and inocula used for application in rice fields. *Biology and Fertility of Soils*, **4**, 98 – 105.
Roger,P. and Watanabe,I. (1986) Technologies for utilizing biological nitrogen fixation in lowland rice: potentialities, current usage, and limiting factors. *Fertilizer Research*, **9**, 39 – 77.
Roychoudhury,P., Krishnamurti,G.S.R. and Venkataraman,G.S. (1980) Effect of algal inoculation on soil aggregation in rice soils. *Phykos*, **19**, 224 – 227.
Saha,K.C. and Mandal,L.N. (1979) Distribution of nitrogen fixing blue-green algae in some rice soils of West Bengal. *Journal of Indian Society for Soil Science*, **27**, 470 – 477.
Sharma,B.M. and Gupta,R.S. (1983) Effect of algal application on rice yield in Jammu Division. *Phykos*, **22**, 176 – 179.
Sharma,V.K. and Gaur,Y.S. (1981) Nitrogen fixation by pesticide-adapted strains of paddy-field cyanophytes. *International Journal of Ecology and Environmental Science*, **7**, 117 – 122.
Shi,D.-J. and Hall,D.O. (1988) *Azola* and immobilized cyanobacteria (blue-green algae): traditional agriculture to biotechnology. *Plants*, **1**, 5 – 11.
Shuying,L. (1987) Methods for using *Azolla filiculoides* sporocarps to culture sporophytes in the field. In *Azolla Utilization. Proceedings of the Workshop on Azolla Use, Fuzhou, Fujian, China, 31 March – 5 April 1985*. International Rice Research Institute, PO Box 933, Manila, Philippines, pp. 27 – 32.
Singh,A.L. and Singh,P.K. (1987) Comparative study on *Azolla* and blue-green algae dual culture with rice. *Israel Journal of Botany*, **36**, 53 – 61.
Singh,L.J., Tiwari,D.N. and Singh,H.N. (1986) Evidence for genetic control of herbicide resistance in a rice-field isolate of *Gloeocapsa* sp. capable of aerobic diazotrophy under photoautotrophic conditions. *Journal of General and Applied Microbiology*, **81**, 81 – 88.
Singh,P.K. and Singh,A.L. (1983) Comparative studies on *Azolla* and blue-green algae biofertilization to

rice crop. In Shukla,A.C. and Pandey,S.N. (eds), *Advances in Applied Phycology*. International Society for Plants and Environment, Kanpur, India, pp. 334–349.

Singh,R.N. (1961) *Role of Blue-green Algae in Nitrogen Economy of Indian Agriculture*. Indian Council of Agricultural Research, New Delhi.

Subrahmanyan,R., Relwani,L.L. and Manna,G.B. (1965) Fertility build-up of rice field soils by blue-green algae. *Proceedings of the Indian Academy of Sciences, Series B*, **62**, 252–277.

Sundara Rao,W.V.B., Goyal,B.S.K. and Venkataraman,G.S. (1963) Effect of inoculation of *Aulosira fertilissima* on rice plants. *Current Science*, **32**, 366–367.

Tirol,A.C., Roger,P.A. and Watanabe,I. (1982) Fate of nitrogen from a blue-green alga in a flooded rice soil. *Soil Science and Plant Nutrition*, **28**, 559–569.

Venkataraman,G.S. (1972) *Algal Fertilizers and Rice Cultivation*. Today & Tomorrow's Printers & Publishers, New Delhi.

Venkataraman,G.S. (1981) *Blue-green Algae for Rice Production. A Manual for its Promotion*. FAO Bulletin No. 46, FAO, Rome.

Watanabe,A., Ito,R. and Konishi,C. (1951) Effect of nitrogen-fixing blue-green algae on the growth of rice plants. *Nature*, **168**, 748–749.

Watanabe,A. (1959) Distribution of nitrogen-fixing blue-green algae in various areas of south and east Asia. *Journal of General and Applied Microbiology*, **5**, 21–29.

Watanabe,A. (1962) Effect of nitrogen-fixing blue-green alga *Tolypothrix tenuis* on the nitrogenous fertility of paddy soil and on the crop yield of rice plants. *Journal of General and Applied Microbiology*, **8**, 85–91.

Watanabe,A. (1973) On the inoculation of paddy fields in the Pacific area with nitrogen-fixing blue-green algae. *Soil Biology and Biochemistry*, **5**, 161–162.

Watanabe,I. (1987) Summary report of the *Azolla* program of the International Network on Soil Fertility and Fertilizer Evaluation for Rice. In *Azolla Utilization. Proceedings of the Workshop on Azolla Use, Fuzhou, Fujian, China, 31 March–5 April 1985*. International Rice Research Institute, PO Box 933, Manila, Philippines, pp. 101–108.

Watanabe,I. and Berja,N.S. (1983) The growth of four species of *Azolla* as affected by temperature. *Aquatic Botany*, **17**, 175–185.

Watanabe,I., Lapis,H.T., Oliveros,R. and Ventura,W. (1988) Improving phosphate fertilizer application to *Azolla*. *Soil Science and Plant Nutrition* (in press).

Watanabe,I., Subudhi,B.P.R. and Aziz,T. (1981) Effect of neem cake on the population and nitrogen fixing activity of blue-green algae in flooded soil. *Current Science*, **50**, 937–939.

Watanabe,I. and Ventura,W. (1982) Nitrogen fixation by blue-green algae associated with deepwater rice. *Current Science*, **51**, 462–465.

Whitton,B.A. (1987) Survival of blue-green algae. In Henis,Y. (ed.), *Survival and Dormancy of Microorganisms*. Wiley-Interscience, New York, pp. 109–167.

Whitton,B.A., Aziz,A., Kawecka,B. and Rother,J.A. (1989) Ecology of deepwater rice-fields in Bangladesh. 3. Associated algae and macrophytes. *Hydrobiologia* **169**, 31–42.

Xiao Qing-yuan (1987) The economic value and use of red duckweed *Azolla* sp. in Anhui Province, China. In *Azolla Utilization. Proceedings of the Workshop on Azolla Use, Fuzhou, Fujian, China, 31 March–5 April 1985*. International Rice Research Institute, PO Box 933, Manila, Philippines, pp. 268–269.

Xiao Qing-yuan, Shi Yan-ru, Yang Guang-li and Peng Ke-lin (1987) Germination of *Azolla filiculoides* Lam. sporocarps and factors affecting their growth. In *Azolla Utilization. Proceedings of the Workshop on Azolla Use, Fuzhou, Fujian, China, 31 March–5 April 1985*. International Rice Research Institute, PO Box 933, Manila, Philippines, pp. 33–45.

CHAPTER 8

The use of engineered and genetically distinct inocula on plants

JOHN E.BERINGER[1], KEITH A.POWELL[2] and NIGEL J.POOLE[2]

[1]*Microbiology Department, Medical School, University Walk, University of Bristol, Bristol BS8 1TD, UK* and [2]*ICI Agrochemicals, Jealotts Hill Research Station, Jealotts Hill, Bracknell, RG12 6EY, UK*

Microorganisms involved in crop nutrition

A number of microorganisms involved in plant nutrition have already been discussed in this volume (Eaglesham, Chapter 3; Stribley, Chapter 4). The importance of many of these is clear, and there are obvious targets for genetic manipulation relating to their ability to provide nutrients. However, as well as improving existing functions, it is useful to speculate about other valuable roles for these microorganisms.

One of the most difficult aspects of pest and disease control is to tackle diseases of roots (Campbell, Chapter 5). This is because most useful chemicals are not transported readily to roots to produce systemic toxicity, and it is undesirable to add large quantities of persistent chemicals to soil to provide control during the growing season. A possible solution to these problems is to introduce appropriate toxin genes into root-colonizing microorganisms and inoculate the seed or soil. An example of this approach is the work of scientists at Monsanto who are interested in inserting the *Bacillus thuringiensis* delta endotoxin gene into a root-colonizing pseudomonad to control the larvae of some insect species. The same idea has been suggested for *Rhizobium* to enable it to produce root nodules resistant to the larvae of *Sitona*, which can cause serious losses of nitrogen-fixing nodules.

Given sufficient ingenuity, there is no reason why genes producing compounds capable of inhibiting nematodes, fungi, bacteria and other root pests should not be exploited (Macdonald, Chapter 1). As with other aspects of gene manipulation, the major problem is usually not the ability to manipulate the microorganisms genetically but our knowledge of the biochemistry and genetics needed to isolate appropriate genes to introduce into them. In this respect recent advances in cloning antibiotic production pathways offer hope that suitable genes will become available in the next decade.

Because they are ubiquitous and have wide host ranges, mycorrhizal fungi (Stribley, Chapter 4) are good candidates for genetic manipulation and there is much interest in exploiting these organisms. For the culturable ectomycorrhizal species this should not present an insuperable problem, because methods for introducing genes into fungi are improving constantly. While many of the methods will be relevant to endomycorrhizal fungi, our inability to culture these species makes it unlikely that they can be exploited in the near future (Hirsch, 1986; Burggraaf and Beringer, 1988).

There are two major problems in exploiting microorganisms that interact with plants. The first is the problem of introducing an organism into an environment in which there is an indigenous population of the same species. The second is the requirement that the introduced microorganism be able to compete for sites on host plants that will exclude the other microorganisms. Almost a century of experience with *Rhizobium* inoculants has shown that competition from indigenous strains makes it very difficult to obtain even as much as 10% of nodules containing the inoculant. At this level of interaction a genetically manipulated strain would need to produce some form of systemic activity within a root system for it to become immune to a target pest species. We know so little about the characteristics needed for an organism to be a successful colonizer and competitor that it is not possible at present to construct 'improved' strains. However, the construction of such strains would, inevitably, lead to problems in obtaining approval for their release (see below).

Agrobacterium

Studies on the control of crown gall disease by *Agrobacterium* were first reported by New and Kerr (1972) who used a non-tumorigenic *Agrobacterium* strain (K84) (Campbell, Chapter 1). Htay and Kerr (1974) showed that K84 controlled the disease in the field, and provided the basis for a large international business (Moore and Warren, 1979; Kerr and Tate, 1984). Control operates largely through the killing of tumour-forming strains by a bacteriocin, agrocin K84 (*Figure 1*), which is taken up into susceptible strains by a permease carried on their tumour-forming (Ti) plasmid (Engler *et al.*, 1975). Not surprisingly, strain K84 is not effective in all soils, because not all pathogenic agrobacteria have uptake permease genes associated with their Ti plasmids (Moore and Warren, 1979), and there is always the possibility of mutation to K84

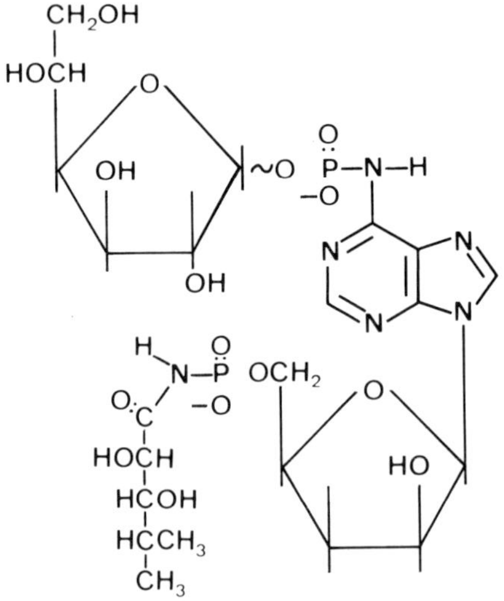

Figure 1. The structure of Agrocin K84, based on Shim (1987).

resistance, which can occur at high frequency in the laboratory (Kerr and Htay, 1974), although this appears not to be a problem in the field (Shim *et al.*, 1987).

Another threat to the use of strain K84 as a control agent is the possibility that resistant pathogens would be produced as a result of conjugation with K84 and transfer of the plasmid carrying resistance and production genes. Shim *et al.* (1987) and Shim (1987) have approached this problem by constructing transfer-deficient (Tra$^-$) mutants of the K84 plasmid which, as expected, were unaffected in bacteriocin production. The construction and use of Tra$^-$ deletion strains of *A. radiobacter* strain K84 for commercial use can be expected in the near future (Shim *et al.*, 1987).

Because it is generally assumed that biological control is mediated through the production of the K84 bacteriocin, it appears logical to improve biological control through the construction of strains producing larger amounts of bacteriocin. Surprisingly, there appears to be no correlation between quantity of bacteriocin and gall control— Shim *et al.* (1987) found that strains producing about 5 times more bacteriocin were no more efficient than the parent strain. When the overproducing strain was a derivative of another *Agrobacterium* strain, control was again less efficient. This may be due to the importance of physical site blockage, as well as bacteriocin production, in disease control (Cooksey and Moore, 1982). The implication of this work must be that the colonization of the treated rootstock and interaction between bacteria are strain-dependent and are an important component of the control system.

Recently Shaw and colleagues (Ashby *et al.*, 1987) have suggested that *Agrobacterium* can be used to 'target' biological control to wounds on plants. The intention is to take advantage of the observation that genes for tumour induction are induced by exudates from wounds (Ashby *et al.*, 1987) and use these regulatory sequences so that agrobacteria are attracted to wounds and then express their cloned toxin genes to kill pests only at the wound. Such target-specific gene expression methods are of great value, because the production of new compounds by organisms can be controlled and exploited for specific purposes.

Fungal pathogens of insects

In 1888 Krassiltchik published a paper describing the use of *Isaria destructor* for control of beet weevil (*Cleonus punctiventris*) on sugar beet. In a semi-industrial process he produced about 200 g spores m^{-2} of surface culture. Spores (8 kg ha^{-1}) were used to achieve '55.80% insect mortality'. He concluded that biological control had become a practical method which in the future would be perfected and used on a large scale. In reality, 100 years on, have we achieved Krassiltchik's predicted success? The answer must be no; there is little use in the Western world of insect pathogenic fungi for biological control.

Several hyphomycete genera are of interest. *Verticillium lecanii* has been used for whitefly control in the UK (Vertalec, Tate & Lyle); *Beauveria bassiana* is reportedly used in Russia (Boverin); *Metarhizium* has been reported to give control of plant hoppers in rice, and *Paecilomyces* is produced as Biocon in the Philippines. However, these examples can hardly be claimed to represent widespread use. Why has so little success been found? There are a number of problems which might be solved through the use of genetic manipulation but only when we have a much better understanding of the biochemistry and physiology of the fungi concerned.

It is often difficult to achieve large-scale production to give viable spore or mycelial preparations: the resultant products are, therefore, more expensive than chemical alternatives. Many fungal spores are difficult to formulate and maintain in a viable state for field use: it is necessary to achieve close to 90% viability for at least 1 year to produce a commercial product. Often field trial results give less than complete control because fungi require high relative humidity for germination and sporulation, do not sporulate sufficiently for rapid spread of infection and do not kill quickly enough to prevent crop damage.

Bacterial pathogens of insects

A number of *Bacillus* species have been examined as potential biological control agents, for example, *B.thuringiensis*, *B.cereus*, *B.popilliae*, *B.laterosporus* and *B.sphaericus* (Aronson *et al.*, 1986). By far the most successful is *B.thuringiensis* (Bt), and since 1962 several thousand tonnes have been used to control certain lepidopteran and dipteran pests (such as *Chloristoneura fumiferana*, *Plutella xylostella* and *Leptinotarsa decemlineata*), without raising any serious concerns about environmental or toxicological side-effects.

Bt is a complex species which is divisible into at least 20, and arguably many more, varieties (Aronson *et al.*, 1986). During sporulation a large parasporal proteinaceous crystal is formed which contains the insecticidal δ-toxin. The crystal is atoxic until solubilized in an alkaline insect gut. The commercial strains of Bt contain both the spore and the crystal, although some asporogenous Bt strains are being commercialized. Some strains (not the commercial varieties) of Bt also produce a β-toxin, called thuringiensin (*Figure 2*). Thuringiensin is a structural analogue of ATP which is being developed as a chemical insecticide.

Bt varieties differ both in the quantity of toxin produced within the crystal and their spectrum of activity against insect species. In 1970 a strain HD1 was isolated from a colony of *Pectinophora gossypiella* (pink bollworm). This strain quickly became the standard commercial strain for the control of lepidopterous pests in both agriculture and forestry. In 1976, another strain, *B.thuringiensis* var. *israelensis*, was discovered to have activity against dipteran pests and has been successfully commercialized for the control of pests such as blackfly (*Simulium damnosum*) and mosquitoes (*Culex*

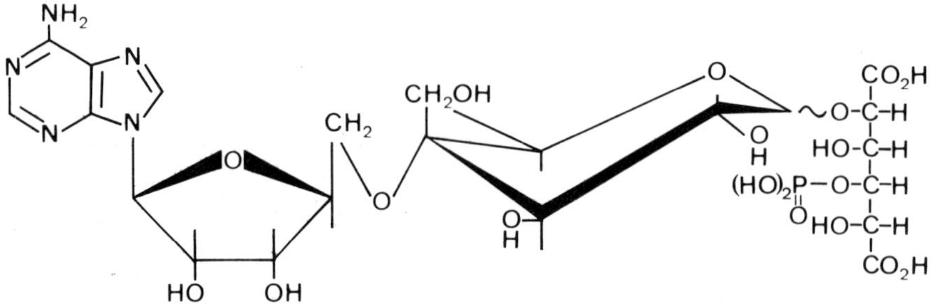

Figure 2. The structure of the β-exotoxin from *Bacillus thuringiensis*.

tarsalis). In the last few years, varieties with activity against coleoptera have been discovered, and one is now in the process of commercialization to control Colorado potato beetle. Bt toxin, like all chemical insecticides, has characteristics which limit its effectiveness. Some of these cannot be corrected; for example, its mode of action is via the stomach and therefore it has to be eaten, although for many pests a contact mode of action is required. However, for other innate characteristics of the toxin it is possible to envisage solutions, for example, increasing the speed of kill, reducing sensitivity to ultraviolet light or widening the spectrum of activity. The latter is crucial, since for many crops it is essential to control a complex of insect pests rather than just one insect species. One solution is, of course, to look for new Bt varieties which are fast-acting or are active against a wider spectrum of pests. Another, possibly complementary, approach is to use genetic engineering techniques.

Gonzalez *et al.* (1981) were able to demonstrate that the structural genes responsible for the crystalline-toxin production are located on high-molecular-weight plasmids in most varieties of Bt. These plasmids are often of different sizes in different varieties. In 1982, Gonzalez *et al.* reported that the plasmid containing the toxin genes and other plasmids can be transferred among some varieties of Bt and *B. cereus* by a conjugation-like process with high efficiency. This has naturally led to the development of novel hybrid Bt varieties using conjugation. For example, recently conjugation was used to produce a hybrid cell, containing the plasmid with the gene coding for the endotoxin with lepidopteran activity from Bt *kurstaki* and the plasmid with the gene coding for coleopteran activity from Bt *tenebrionis* (Karamata and Piot, 1987).

Genetic manipulation into pseudomonads

Two notable examples exist of the use of *Pseudomonas* species in genetic manipulation. Mycogen have used *Pseudomonas* as a host for cloned toxin genes from *B. thuringiensis*. The purpose of this approach has been described as the microbial encapsulation of toxin protein. The resultant preparation is killed to provide a dead cell. It is claimed that this preparation produces a more stable formulation, resistant to environmental causes of toxin degradation, ultraviolet light and dehydration.

Monsanto have isolated a *P. fluorescens* strain which is claimed to colonize roots very efficiently and hence has the potential to provide a method of delivery (of products produced by cloned genes) to plant roots. A rifampicin-resistant isolate has been shown to be as capable of colonizing roots as sensitive isolates, and has been transformed to be able to metabolize lactose, by adding the *lacZ* gene (Drahos *et al.*, 1986), a property rare among fluorescent pseudomonads. The objective of this construction was to enable rapid and sensitive detection of the organism in soil. It is claimed that less than 10 colony-forming units per g of soil can be detected by this method (Drahos *et al.*, 1986). Survival, intact, of the plasmid carrying the *lacZ* gene after growth of the bacteria for 3–4 weeks on soyabean roots was reported.

The genetics studies done to date have clearly been extensive and have provided clear potential for the introduction of foreign genes into soil bacteria. It should be a small step from the work described to the commercial production of a live organism with foreign genetic material. As yet, no clear view exists of the regulatory position with respect to such bacteria, but clearly the Monsanto experiments will provide useful data to allow an informed debate on the merits of release of these bacteria.

Microbial weed control

Many microbial herbicides have been developed, and two products exist: Collego for control of northern joint-vetch (*Aeschynomene virginica*) in rice in the USA and Devine for control of milkweed vine (*Morrenia odorata*) (Lethbridge, Chapter 2). Both of these products are based on the use of living fungi for the activity. A major problem for the exploitation of microorganisms for weed control has often been the establishment of the first stage of infection. The fungi require a dew period for infection, and this frequently may not be available in major dry-land crops. It is perhaps not surprising the Collego is sold for a rice weed killer. A secondary problem is the sensitivity of the fungi used to fungicides used in treatment of the crop for fungal diseases.

Major improvements for microbial herbicides must depend on overcoming the initial dew-period requirement. It is doubtful that such an improvement could be obtained by genetic means. It may be possible to look for novel toxophore production by genetic transfer; for example, the ability to produce Bialophos, a microbially produced herbicide, could be transferred *but* it would not enhance activity, in the absence of infection. A second problem is the desirability of producing microbial herbicides which are specific for one, or only a few, weeds to reduce the potential for harm to other plants in the environment. Given that there are few situations where a crop is infested with a single weed, there are few opportunities for a high-cost single-weed product. This, coupled with the biological problems associated with establishing the fungi, make it unlikely that genetically manipulated products will be common in the foreseeable future.

Microbial fungicides

Although microbial fungicides have been proposed there are, as yet, only two products for field crops. Dagger, a pseudomonad, is used to control *Pythium*, and Quantum 4000, a *Bacillus subtilis* strain, is used to control peanut diseases. Given that these two products can provide some degree of microbial colonization, then there may be some potential for genetic improvement. An example of the type of added activity which might be valuable is the chitinase from *Serratia marcescens* (Jones *et al.*, 1986). If this were added to an effective colonizing agent, it might provide augmented activity.

The major problem in this area is the need for novel activity which could be of value. It may again require extensive research to enable the discovery of additional factors giving increased activity against plant pathogens.

Plant growth-stimulating bacteria

If we ignore the bacteria which stimulate plant growth by giving some degree of control of pathogens (Campbell, Chapter 1) or those which fix nitrogen symbiotically (Eaglesham, Chapter 3), we are left with those like *Azospirillum* which are claimed to stimulate plant growth by hormone products or associative nitrogen fixation. We have shown with wheat (Powell and Lethbridge, 1987) that there is no significant effect of *Azospirillum* on yield, and remain to be convinced that other bacteria are capable of giving this effect, whether or not they are manipulated genetically.

Conclusion

A great deal needs to be done to understand the interaction of microbial products with pathogens, weeds and insects before genetic manipulation can offer a complete answer to the lack of activity. We should be directing the effort in genetic manipulation to an understanding of what makes some microbes active in biological control, and to addressing the requirement for improved control, before random approaches are likely to be successful.

The regulatory position

We should see genetically manipulated microorganisms entering the marketplace by the turn of the century to improve nutrition (for example, *Rhizobium*) and for biological control (for example, *Pseudomonas*). Before such organisms can be marketed there will be a need to obtain approval for their release into the environment. Because microorganisms are very small, are difficult to isolate when present in very small numbers in soil, and cannot be recalled after release, it will be necessary to convince regulatory bodies that strains with new or improved abilities are safe and do not have the potential to cause harm in the environment. It is not clear yet how difficult it will be to obtain approval for releases in the future, because of the difficulty of presenting the case that organisms will not cause harm.

Experience with the 'ice-minus' organisms (Campbell, Chapter 5) and Monsanto's Lac^+ pseudomonad in the USA indicates that organisms that are not used as inoculants at present will require fairly extensive documentation and will be expensive to regulate. By comparison, *Rhizobium* should be simpler because the release of strains is widespread and has occurred for a long time. The recent experience in the USA of Biotechnica International has shown that, for genetically manipulated rhizobia, approval was not too difficult to achieve. It is to be hoped that, as experience builds up and knowledge of microbial ecology improves, so also will our ability to assess risks.

One of the most difficult aspects of regulation involves the conflict between the need for inoculants to become well established and be competitive, and the regulatory requirement that organisms have not been manipulated to increase their potential to become pests. Perhaps the only solution will be to manipulate strains so that they are no longer persistent through the introduction of genes conferring conditional lethality or reduced vigour. For this to be possible, we will need a much better understanding than we have at present of the factors involved in persistence in the environment and the ability to interact with host plants.

References

Aronson,A.I., Beckman,W. and Dunn,P. (1986) *Bacillus thuringiensis* and related insect pathogens. *Microbiological Reviews*, **50**, 1−24.

Ashby,A.M., Watson,M.D. and Shaw,C.H. (1987) A Ti-plasmid determined function is responsible for chemotaxis of *Agrobacterium tumefaciens* towards the plant wound product acetosyringone. *FEMS Microbiology Letters*, **41**, 189−192.

Burggraaf,A.J.P. and Beringer,J.E. (1988) Absence of nuclear DNA synthesis in vesicular-arbuscular mycorrhizal fungi during *in vitro* development. *The New Phytologist* (in press).

Cooksey,D.A. and Moore,L.W. (1982) Biological control of crown gall with an agrocin mutant of *Agrobacterium radiobacter. Phytopathology*, **79**, 919−921.

Drahos,D.J., Hemming,B.C. and McPherson,S. (1986) Tracking recombinant organisms in the environment: β-galactosidase as a selectable non-antibiotic marker for fluorescent pseudomonads. *BioTechnology*, **4**, 439−444.

Engler,G., Holsters,M., van Montagu,M., Schell,J., Hernalsteens,J.P. and Schilperoort,R. (1975) Agrocin 84 sensitivity: a plasmid determined property of *Agrobacterium tumefaciens. Molecular and General Genetics*, **138**, 345−349.

Gonzalez,J.M., Dalmage,H.T. and Carlton,B.C. (1981) Correlation between specific plasmids and δ-endotoxin production in *Bacillus thuringiensis. Plasmid*, **5**, 351−365.

Gonzalez,J.M., Brown,B.J. and Carlton,B.C. (1982) Transfer of *Bacillus thuringiensis* plasmids coding for δ-endotoxin among strains of *B.thuringiensis* and *B.cereus. Proceedings of the National Academy of Sciences of the USA*, **79**, 6951−6955.

Hirsch,P.R. (1986) Gene cloning and its potential application to mycorrhizal fungi. In Gianinazzi-Pearson,V. and Gianinazzi,S. (eds), *Physiological and Genetical Aspects of Mycorrhizae*. INRA, Paris, pp. 147−157.

Htay,K. and Kerr,A. (1974) Biological control of crown gall: seed and root inoculation. *Journal of Applied Bacteriology*, **37**, 525−530.

Jones,J.D.G., Grady,K.L., Suslow,T.V. and Bedbrook,J.R. (1986) Isolation and characterization of genes encoding two chitinase enzymes from *Serratia marcescens. The European Molecular Biology Organization Journal*, **5**, 467−473.

Karamata,D. and Piot,J.C. (1987) Hybrid *Bacillus thuringiensis* cells useful for biological control of pests. European Patent Application J8610469.6.

Kerr,A. and Htay,K. (1974) Biological control of crown gall through bacteriocin production. *Physiological Plant Pathology*, **4**, 37−44.

Kerr,A. and Tate,M.E. (1984) Agrocins and the biological control of crown gall. *Microbiological Sciences*, **1**, 1−4.

Krassiltchik,J.M. (1888) La production industrielle des parasites végétaux pour la destruction des insectes nuisibles. *Bulletin Scientifique de la France et de la Belgique*, **19**, 461−472.

Moore,L.W. and Warren,G. (1979) *Agrobacterium radiobacter* strain 84 and biological control of crown gall. *Annual Review of Phytopathology*, **17**, 163−169.

New,P.B. and Kerr,A. (1972) Biological control of crown gall: field measurements and glasshouse experiments. *Journal of Applied Bacteriology*, **35**, 279−287.

Powell,K.A. and Lethbridge,G. (1987) The investigation of benefit from *Azospirillum* species in agriculture. *Journal of the Science of Food and Agriculture*, **40**, 121−122.

Shimm,J.-S. (1987) Studies of the agrocin 84 plasmid of *Agrobacterium radiobacter*. PhD Thesis, University of Adelaide, Australia.

Shimm,J.-S., Farrand,S.K. and Kerr,A. (1987) Biological control of crown gall: construction and testing of new biocontrol agents. *Phytopathology*, **77**, 463−466.

INDEX

Abutilon, 84
Acer, 57
Activity spectrum of CP & E agents, 22
Adsorption, sites for on roots, 69
Adzuki, 33
Aeschynomene, 5, 8, 14, 16, 22, 36, 106
   virginica, 14, 106
Africa, *Rhizobium* in, 31, 34
Agriculture, practices in,
   and biocontrol, 75
   and disease, 74
   and mycorrhiza, 55
Agro-forestry, 2
Agrobacterium, 84, 102
   colonization by limits interaction, 103
   radiobacter, 5
   radiobacter 'strain 84', 69, 102
   tra minus strains, 103
   tumefaciens, 5, 69, 72
   wound inoculation by, 103
Agrochemicals, 11
   industrial production of, 11−14
Agrocin 84, 69, 72, 102
Akinete, 92, 95
Alfalfa mosaic virus, 82
Algalization, 92−93, 97
Alginate, 24
Alley cropping systems, 31
Alnus, 2
America,
   Rhizobium in central and south, 37
   Rhizobium in north, 36
Anabaena, 2, 89
   azollae, 95
   variabilis, 96
Antagonism, biocontrol and, 17
Antagonistic microorganisms, *see* Microorganisms
Antagonists, 67, 72
   to soil-borne pathogens, 14
Antibiosis, 3
Antibiotic, 17, 69, 71−72
   manipulation of genes for, 101
   super-producers of, 103
Antisense RNA, 81−82
Aphids, 7
Apple, 5
Apricot canker, 69
Arachis, 41
Arbuscules, 50
Army worms, 6
Arthrobotrys,
   irregularis, 8
   robusta, 8

Ascomycetes, 51
Ash, 57
Asia, *Rhizobium* in, 33, 38, 40
Asparagus, 57
Asporogeny, *Bacillus thuringiensis* strains and, 104
Associative fixation, 2
Astralagus, 33
Attachment sites on host, 72
Aulosira, 89, 93
Australia, *Rhizobium* in, 32
Autotrophic host, 49
Azadirachta indica, 93
Azolla, 2, 89−100
   as animal feed, 94
   culture of, 94
   effect of competition on, 95
   environmental factors and, 95
   filicoides, 95
   water availability and, 95
Azospirillum, 2, 106
Azotobacter, 1−2
Azotobacterin, 1, 25

Bacillus,
   cereus, 104−105
   laterosporus, 104
   megatherium, 1
   popilliae, 6−7, 104
   radicicola, 29
   sphericus, 104
   subtilis, 106
   thuringiensis, 3−7, 14, 17−22, 26, 101, 104
      for beetle control, 105
      global market for, 18
      selling price of, 21
      toxins of, *see* Bt toxin
      value of, 18
      var. *israeliensis,* 5, 104
      var. *kurstaki,* 105
      var. *tenebrionis,* 105
Bacteria,
   as biocontrol agents, 3
   deleterious, 70
   as insect control agents, 3
   as pathogens of insects, 4, 104
   plant disease and, 101
   symbiosis of with nematodes, 7
Bacterial brown blotch, *see* Mushroom
Bacteriocin, 69, 72, 102
   super-producers of, 103
Bambara, 31
Barley, 55, 57

Index

Basidiomycetes, 51
Bean, 31–37, 40, 44
Beaveria bassiana, 103
Beet, 51
Beet weevil, 103
Benzimidazole fungicides, 14
Berseem, 35, 39
Beta endotoxin, *see* Bt toxin
Bialophos, 106
Binab T, 5
Biobit, 19
Biocon, 103
Biocontrol (*see also* Control, biological)
   agents for, 3, 101
   by *Agrobacterium*, 103
   commercial, 67
   future of, 73, 106–107
   fungal, 3
   of diseases, 67–77
   of fungi, 106
   of insects, 3, 103
   of weeds, 14, 106
   screening for agents, 14
Biorational pesticides, *see* Pesticides
Black gram, 39
Blackfly, 5, 18, 104
Blitox, 93
Blue-green algae, 89–100 (*see also* Cyanobacteria)
   availability of water and, 95
   future of, 96–97
   germination inhibition by, 90
   germination promotion by, 90
   symbiotic associations and, 96
Bollworm, 104
Boverin, 103
Bradyrhizobium, 43, 45
   fermentation cost of, 21
   japonicum, 18, 34
Bran, 24
Brazil, *Azolla* culture in, 94
Brassica, 5
Broad bean, 32, 37, 39
Brown rice hoppers, 103
Bt toxin, 4, 6, 17, 101, 104–105
   gene for in plants, 101
   plasmids containing gene for, 105
   transferred to *Pseudomonas*, 105
Buckwheat, 29
Butt rot of conifers, 67, 69

Cabbage moth, 5–6
Cadmium, 54
Calliandra, 31
Calopogonium, 40

Calothrix, 90, 96
Caragana, 33
Carbaryl, 93
Carbon competition, 72
Carribean, *Rhizobium* in, 32
Cassava, 57–58
Cassia, 31
Casuarina, 2
Cauliflower mosaic virus, 85
Caulimoviruses, 84
Cecids, 5
Central America, *Rhizobium* in, 32
Centro, 32, 37, 40
Centrosema, 41
Ceratocystis ulmi, 5
Cereal, 73
Certan, 6, 19
Chelating agents, iron and, 70
Chemical control of disease, *see* Control
Chemically induced resistance, *see* Resistance
Chenopodiaceae, 50
Chickpea, 32–39
China, *Azolla* culture in, 94
Chitinase, 83, 106
Chlamydospores, mycorrhizal, 50
Chloristoneura fumiferana, 104
Chlorogloea fritschii, 96
Chlorogloeopsis fritschii, 96
Chondrostereum purpureum, 69
Citrus, 23, 57, 80
Citrus tristeza virus, 80
Cleonus punctiventris, 103
Cloning of antibiotic genes, 101
Clover, 31–39, 57
Coccineus bean, 34
Codling moth, 5
Coffee, 57
Coleoptera, 7
   control of by *B.thuringiensis*, 105
Collego, 5, 8, 14, 16, 19, 22–23, 25, 106
   fermentation cost of, 21
Colletotrichum gloeosporioides, 5, 14, 19
Colocasia, rice culture and, 94
Colonization, 6
   ability for, 74, 106
   by *Agrobacterium*, 103
   by mycorrhiza, 55, 58
   characteristics of, 102
   of antagonist, 74
   of roots, 55, 68, 101, 105
Colorado beetle, control of, 105
Commercial considerations, *see* Biocontrol; Fungi; CP & E agents; *Rhizobium*; Viability

Index

Commodity product, 12, 20
Competition,
  between *Rhizobium* strains, 45, 102
  for attachment sites, 72, 102
  for carbon, 72
  for nutrients, 71
  of blue-green algae, 93
  with *Azolla*, 95
  with introduced inocula, 68
Competitive ability of inocula, *see* Inocula; *Rhizobium*
Competitive exclusion, 69
Complementary RNA, 81
Compost, 68, 73
Conditional lethality, 107
Coniothyrium, 73–74
Control,
  biological, 3, 67–77 (*see also* Biocontrol)
  chemical, 68
  of pests and diseases, 1
Control agents,
  commercial, 67
  costs of, 6, 104
  discovery of, 4, 13
  testing of, 14–17
Controlled environments, 26
Copper, 53–54
Cost analysis, 73
Cost:benefit ratio, 16, 20
Costs,
  of control agents, 6, 104
  of development, 13
  of fermented products, 21
  of microbial herbicides, 106
  of *Rhizobium* inoculum, 41
Cotton, 16, 22
Cotton bollworm, 7
Cover crops for rubber, 40
Cowpea, 31–40, 44
CP & E, *see* Crop protection and enhancement
Crop protection and enhancement, 11–28
  agents for, 21, 22, 23, 26
  application of, 23
  commercial use of, 11–25
  development of market for, 12
  discovery of, 13
  future of, 26
Crop protection, microbial, 11
Cropping practices and mycorrhiza, 60
Crops,
  management of, 67
  nutrition of, 1
  rotation of, 71, 74

  sequence of, 74
  yield of, 30–41
Cross protection, 80–82
  by virus, 79
Crotalaria, 37, 39
Crown gall, 5, 69, 72, 102–103
Cruciferae, 50
Cucumber mosaic virus, 81
Culex tarsalis, 104
Culicidae, *see* Mosquito
Cultural practices, mycorrhiza and, 55
Customer demand, 20
Cyanide, production of in soil, 71
Cyanobacteria, 2, 89–100 (*see also* Blue-green algae) nitrogen-fixing, 89
Cyanogenic glycosides, 71
Cycas, 2, 96
Cydia pomonella, 5
Cylindrospermum, 90

Dagger, 106
Damping off, 68–69
Deepwater rice, 91, 96
Delta endotoxin, *see* Bt toxin
Depletion, zone of around roots, 53
Desert irrigation for *Azolla*, 95
Desiccation of blue-green algae, 92
Desmodium, 33, 41
Devine, 19, 23, 106
Dew period (*see also* Humidity, relative)
  and fungal inocula, 106
Dipel, 5, 19
Diptera, control by *Bacillus*, 104
Discovery
  costs, 14
  of control agents, 4
  of CP & E agents, 13
Disease,
  control of, 67, 101
  protection against, 54
  resistance to, 3
  tolerance of, mycorrhiza and, 58
Disease-suppressive soil, 14
DNA, 84
Drought, tolerance of, 3
Drugs, 1
Dutch elm disease, 5

Ecology, microbial, 27, 67
Ectomycorrhiza, 50–54
  gene manipulation in, 101
  in forestry, 56
  inoculum production of, 58
  market size for, 61
Efficacy,

111

Index

of Bt toxin, 105
of control agents, 4, 21, 74
of CP & E agents, 21–22
of fungal inoculum, 104
of inoculum, 33, 43
of mycorrhiza, 56
of *Rhizobium*, 32, 36, 39, 41, 43
of virus inoculation, 80
efficiency of fertilizer use, 58, 60
Ehrharta calycina, 53
Elcar, 19
Elm, 5
End user price, 20
Endogonaceae, 51
Entomogenous fungi, 3, 103
Entomophthorales, 7
Environment,
  effects of control agents on, 104
  protection of, 67
  safety of, 20
Environmental impact, 4
Environmental Protection Agency (EPA), 14
  registration data for, 15
EPA, *see* Environmental Protection Agency
Ephestia kuhniella, 3
Ericaceae, 51
Ericoid mycorrhiza, *see* Mycorrhiza
Erwinia amylovora, 69
  herbicola, 69
Europe, *Rhizobium* in, 34
Eutypa armeniaceae, 69

Faba bean, 31–36
Fagaceae, 51
Fenugreek, 35
Fermentation, 17, 20
  conditions for, 6
  cost of, 20–21
Fermenter-adapted strain, 75
Fertilizer, 30
  bacterial, 1
  microbial, 1, 11, 18, 25
  nitrogen, 42
Field conditions, 14
Field trials, 74
  for fungal inocula, 104
  for *Rhizobium*, 39
  of mycorrhiza, 56
  reliability of, 21
  variability of, 75
Fine chemicals, 1
Fire blight, 68–69
Fischerella, 90

Flagella antigens, 4
Flavourings, 1
Flemingia, 31
Flower break by virus, 84
Fluorescent *Pseudomonas*, 69, 72, 105
Foliar pathogen, 68
Fomes, *see Heterobasidion*
Forest nurseries, 56
Forestry and biocontrol agents, 18
Formulation, 7, 20, 24
  of control agents, 23
  of fungal inocula, 104
  of inoculum for ectomycorrhiza, 58
  of mycorrhiza inoculum, 57
  of *Rhizobium* inoculum, 30–42
Frankia, 2
Fraxinus, 57
Frost damage, 69, 71
Fruit and vegetables, 16
Fumigation of soil, 73
Fungi
  commercial production of, 104
  control of pests by, 7
  in biocontrol, 3, 101
  large-scale production of, 104
Fungicides, 12, 14, 68
  market value of, 16
  microbial, 22, 106
  resistance of antagonists to, 69
Fusarium, 69
  lateritium, 69
  moniliforme, 54
  oxysporum, 54, 72
    f. sp. *cucumerinum*, 70
    f. sp. *lini*, 70
  siderophore of, 70
Fusarium-suppressive soils, 72
Fuschia, 57
Future, developments in, *see* Biological control; Blue-green algae; Crop protection and enhancement; Genetic engineering; Herbicides; Mycorrhiza; Plant growth promotion; *Rhizobium*; Virus inoculants

Gaeumannomyces graminis, 69
Galltrol, 5
Gene vectors, 84
Genetic engineering, 3, 5–6, 24–25, 30, 67, 71, 74, 82, 84, 97, 101–108
  and release of organisms, 105
  and safety, 107
  future of in biocontrol, 105
  of *Bacillus thuringiensis*, 105
  of *Rhizobium*, 44

of viruses, 24
Genetic manipulation, 101–108
  of *Pseudomonas*, 105
Genetic variation in mycorrhiza, 56
Genetically distinct inocula, 101–108
Germination, effect of blue-green algae on, 90
Gilpinia hercyniae, 7
Glasshouse screening, 14
Gliricidia, 31, 37, 40
Gloeocapsa, 93
Gloeotrichia, 90
Glucanase, 83
Glycine, 37
Glyphosate, 22
Grafting in virus transfer, 79
Grain legumes, 30–41
Grain yield of rice, 97
Granular inoculum, *see* Inoculum
Grapes, 18
Grasses, 57
Grazing,
  of *Azolla*, 95
  of blue-green algae, 91, 93
Green manure, 74
  and *Azolla*, 94
  and rice, 33
Groundwater, analysis of, 13
Growth hormones, production of, 106
Growth regulators from blue-green algae, 90
Growth, stimulation of, 106
Gunnera, 96

H antigens, 4
Heavy metals,
  and mycorrhiza, 53
  in soil, 54
  tolerance of, 54, 61
Hedysarum, 34–35
Hekiotus armigera, 7
  Heliothis nuclear polyhedral virus, 19
  Herbaspirillum, 2
Herbicides, 12–13
  future of inoculants, 106
  market value of, 16
  microbial, 23, 106
Heterobasidion, 67–69
Heterocystous blue-green algae, 89
Heterorhabditis heliothidis, 5
Heterorhabdus, 7
  Hirsutella thompsonii, 19
Hordeum vulgare, 55
Horsegram, 39
Horse bean, 33

Horticulture, 26
  crops for biocontrol, 73
Host, response of to mycorrhiza, 55
Host specificity of *Rhizobium*, 43
Humidity,
  relative, 5, 7, 23
  and inocula, 104, 106
hup gene, 44
Hydrogen recycle genes, 44
Hymenoscyphus ericae, 56, 59
Hippophae, 2

Ice, damage from, 69, 71
Ice-minus *Pseudomonas*, 107
Ice nucleation, 69
  bacteria and, 71
Identification methods, immunological, 67
India
  Azolla culture in, 94
  rice culture in, 90–100
Indigofera, 36
Indigenous strain competition, 17, 45
Induced protection, 83
  by virus, 79
Induced resistance, 83–84
Ineffective *Rhizobium* strains, 45
Inhibition, zones of on agar, 72
Inhibitor, viral-replication, 83
Inoculation with ericoid mycorrhiza, 59
Inoculum,
  blue-green algal, 89–91
  competitiveness of, 45, 68, 107
  concentration of, 57, 59, 81
  efficacy of, 33, 43, 80, 104
  formulation of, 30–42, 57, 81, 104
  fungal, production of, 104
  granular, 37
  mycorrhizal, 49, 55, 59
  of *Rhizobium*, 29–48
  oil-based, 42–43
  reduced vigour of, 107
  soil-based, 40
Insects,
  fungi pathogenic on, 3, 103
  pathogens of, 3, 103
  pheromones from, 14
  viruses of, 7
Insecticides, 12, 18–19
  market value of, 16
  microbial, 18, 22
Integrated control, 6–7, 68, 73–74
Integrated past management (IPM), 16–17
Intercropping,
  in rice, 94
  with *Azolla*, 94

113

Index

Iron, 53
  phosphates of, 52
Iron-limiting environment, 70
Isaria destructor, 103

Japanese beetle, 6
Juniperus, 57

Klebsiella, 2
Kudzu, 57–58

Lablab, 39
LacZ gene, 105
Lasso, 13
Lathyrus, 39
Lead, 54
Lead time for CP & E agents, 15
Leaf disease, 74
Leek, 57
Legumes, 29–48, 57–58
  as cover crops, 40
  productivity of, 41
Lentil, 32–39
Lepidoptera, 3, 7
  control of by *Bacillus,* 104
Leptinotarsa decemlineata, 104
Lethal genes, 107
Leucaena, 31–32, 40
Ley farming system, 32
Light, requirement of for *Azolla,* 95
Lima bean, 31, 33, 36, 39
Lime, blue-green algae and, 91, 92
Liquid inoculum for *Rhizobium,* 36
Localization of virus infection, 79, 83–84
Long bean, 40
Lotus, 37
Low-input agriculture, 31
  biocontrol in, 75
Lucerne, 31–39
Lupin, 29, 31–37
Luxury uptake of phosphorus, 53
Lysis of hypae, 69

Maize, 16, 57
Malaysia, rice culture in, 90
Mancozeb, 93
Manganese, 53
Maple, 57
Market
  for biocontrol agents, 11
  for CP & E agents, 12, 15, 18
  global, 11, 18
  for microbial herbicides, 106
  size of, 4, 13, 15, 20, 31, 60–61, 73
Market value,

of *Bacillus thuringiensis,* 20
of CP & E agents, 18
of crops, 16
Mechanism,
  cross-protection, 81
  viral action, 83
Medics, 31–32, 35
Melilotus, 36–37
Metals (*see also* Heavy metals)
  toxicity of and mycorrhiza, 53
  uptake of and mycorrhiza, 58, 60
Metarrhizium, 103
Methotrexate, viral resistance to, 85
Mexican yam, 32
Microbes, ecology of, 27, 67
Microbial crop protection, *see* Crop
  protection; Fungicides; Herbicides;
  Insecticides; Pesticides
Microbial fertilizers, *see* Fertilizers
Microorganisms, antagonistic, 21
Mild virus strains, 79–80
Milk weed vine, 23, 106
Mine spoils, mycorrhiza and, 61
Minerals, uptake of, 70
Minor pathogen, 70, 74
Mode of action, 22
  of Bt toxin, 105
  of control agents, 4
  of virus, 81, 83
Models, mathematical, 53
Monterey pine, mycorrhiza and, 56
Montmorillonite clays, 57
Morrenia odorata, 23, 106
Mosquito, 5, 18, 104
Moth, larvae of, 3
Mucana, 37
Mungbean, 32–33, 39–40
Mushroom, 5, 73
  bacterial brown blotch of, 73
Mutualistic symbionts, *see* Symbionts
Mycar, 19
Mycoaid, 59
Mycoherbicide, 8, 14, 19, 21–22
Mycoinsecticide, 19
Mycoparasite, 69, 72–73
Mycorrhiza, 3, 49
  biology of, 50
  colonization by, 53
  disease tolerance and, 58
  ericoid, 50–52, 56, 61
  external hyphae of, 52
  future use of, 61, 101
  gene manipulation in, 101
  inocula from, 49, 60
  mine spoil and, 61

orchids, 50, 54
physiology of, 49
Mucuna, 40
Mycotal, 5, 19

Neem, blue-green algae and, 93
Nematicide, 8
Nematode, 7, 101
Nematophagous fungi, 8
New Zealand, *Rhizobium* in, 32
Niche exclusion, 3
Niche space, 72
Nickel, 54
Nicotiana debneyi, 83
  glutinosa, 83
Nitragin, 29
Nitrogen,
  blue-green algae and uptake of, 90
  fixation of, 2, 29, 89, 106
  requirements of rice for, 89−90
  uptake of, 1, 49, 59, 89, 90
Nitrous acid mutagenesis, 80
Nodulation, root, 2
North America, *Rhizobium* in, 34
Northern joint vetch, *see* Aeschynomene
Nostoc, 90−93, 96
Nursery compost, 68
Nutrients,
  competition for, 71
  uptake of, 49, 101

Oats, 29
Oidendron, 56
Oil-based inoculum, *see* Inoculum
Onion, 52, 57
Operators, safety of, 4
Opines, 72
Orchid, protocorms of, 49
Orchid mycorrhiza, 50, 54
  inoculum for, 59
  market size for, 61
Orchidaceae, 51
Organic farming, mycorrhiza and, 60
Over-producer strains, 103

Paecilomyces, 103
Parasporal protein toxin, 104
Pasture legumes, 30
Patents, 4, 12, 24, 68
Pathogenesis-related proteins, 83−84
Pea, 29, 31−40, 57
Peanut, 31−40
  diseases of, 106
Pear, 5
Peat, 24
  as base of *Rhizobium* inoculum, 32, 37, 40
Pectinophora gossypiella, 104
Pelargonium, 57
Peletons, 50
Peniophora, 68−69
  gigantea, 67, 71
Pentachloronitrobenzene, 73
Pepper, 57, 80
Perlite in *Rhizobium* inoculum, 33
Persistence,
  of control agents, 6
  environmental, 20, 22−23, 107
  of *Rhizobium*, 36
Pest biocontrol agents, 3
Pest management, integrated, 16−17
Pests,
  control of, 3
  of roots, 101
Pesticides, 11
  biorational, 14
  development of, 67
  microbial, 4, 11
  resistance to in blue-green algae, 93
PGPR, *see* Plant growth promoting rhizobacteria; Plant growth promotion
pH,
  blue-green algae and, 90−92
  of soil and siderophore, 70
  Rhizobium and, 34, 36
Phaseolus vulgaris, 31, 44
Phasey pea, 33
Philippines, rice culture in, 90, 94
Phorids, 5
Phosphate,
  as fertilizer, 58−59
  soil, deficient in, 60
Phosphobacterin, 1, 25
Phosphorus,
  blue-green algae and, 90−92
  limiting growth of *Azolla*, 95
  luxury uptake of, 53
  uptake of, 3, 49, 52, 59, 95
Phytophthora, 69
  palmivora, 19, 23
Pigeonpea, 31−32, 37, 39−40
Pinaceae, 51
Pine caterpillar, 7
Pinus radiata, 56
  sylvestris, 54
Pisolithus tinctorius, 59
Plant growth hormones, 70, 106
Plant growth promoting rhizobacteria (PGPR), 70, 74

Index

Plant growth promotion, 1, 3, 11, 70, 90, 106
  future of, 106
Plant growth regulators, 1, 90, 106
Plant hoppers, 103
Plant hormones, 1–2
Plants,
  diseases of, 67
  nutrition of, 1, 101
  productivity of, 68
Plasmid survival for *LacZ* gene, 105
Plasmids,
  of *Bacillus thuringiensis*, 105
  transfer of, 103
Ploughing, mycorrhiza and, 55
Plutella xylostella, 104
Portugal, rice culture in, 90
Potato, 57, 70, 73
Production costs of, 20
Productivity, *Rhizobium* and, 41
Protected environments, 14
Protection, virus-induced, 79
Protoplast fusion, 96
Protoplast test systems, 81
Pseudomonas, 67–68, 70, 74, 106–107
  fluorescens, 69, 72, 105
  genetic manipulation of, 105
  LacZ gene and, 105
  putida, 69–70, 72, 105
  root colonization and, 101
  syringae, 69
  with Bt toxin, 6
Pueraria, 32, 37, 40, 41, 58
Pythium, 69, 73, 106
  ultimum, 54

Quantum 4000, 106

Rape, 29, 51
Recombinant microorganism, 24–25
Registration,
  costs of, 13–14
  for release of inoculum, 107
  of control agents, 73
  of CP & E agents, 14, 24, 107
  of genetically engineered organisms, 105
Release of engineered organisms, 105, 107
Reliability of CP & E agents 21
Replant problems, 71
Research and development, costs of, 4, 13
Residues, 4 (*see also* Persistence)
Resistance,
  of blue-green algae, 93
  to agrocin 84, 102–103
  to biocontrol agents, 7
  to CP & E agents, 17
  to disease, 3, 53
  to pesticides, 14
  to virus, chemically induced, 83
Rhizobitoxin, 45
Rhizobium, 2, 18, 29, 75, 97, 101–102, 107
  commercial inoculants and, 31–41
  competitiveness of inocula of, 45, 102
  efficacy of, 41
  future of, 30, 41
  in different countries, 31–40
  interaction with legumes of, 44
  meliloti, 30, 34, 65
Rhizoctonia, 51, 54, 65
  solani, 73
Rhizopogon vinicola, 54
Rhizosphere,
  population of, 74
  toxin genes of bacteria in, 101
Rice, 5, 8, 16, 23, 106
  culture of, 89
  weeds of, 106
Rice bean, 40
Risk assessment, 25, 73
River blindness, 5
RNA, 84
  viral, 81
Rock phosphate, 52, 58, 60
Root,
  colonization of, 68, 101
    by *Pseudomonas*, 105
  disease of, 68, 74
  nodules, 29, 101
Roundup, 13
Ruderal organism, 74

Safety, 4, 17, 20, 24, 74, 107
  environmental, 17, 20
  operator, 4
Sainfoin, 34, 36
Satellite RNA, 81–82
Sciarids, 5
Sclerotia, 73
Sclerotinia, 69
  minor, 73
  sclerotiorum, 70
Sclerotium, 69
  cepivorum, 73
  rolfsii, 73
Screening,
  for agents, 14, 72
  in vitro, 72
Scytonema, 90
Search for control agents, 72

Index

Seed coat, 69
Seed inoculum for *Rhizobium*, 37
Selection of blue-green algal strains, 93
Selling price of CP & E agents, 21
Serradella, 29, 31–32
Serratia marcescens, 106
Sesbania, 39–40
  rice cultures and, 94
Shelf life, 6, 20, 23
  of *Rhizobium*, 36
Siderophore, 69–72
  –minus mutants, 71
Silver leaf disease, 69
Simulid blackfly, 5
Simulidae, *see* Blackfly
Simulium damnosum, see Blackfly
Siratro, 32–33, 35, 37
Sitona, 101
Skeetal, 19
Soft fruit, 5
Soil-based inoculum, *see* Inoculum
Soil conditioners, 25
Soil type, disease and, 74
Soil-borne diseases, 67, 75
Sorghum, 57
South America, *Rhizobium* in, 37
Soyabean, 5, 8, 16, 18, 23, 31–40, 73, 105
  root colonization of, 105
Specificity of microbial herbicides, 106
Spodoptera, 6
Spores,
  of blue-green algae, 92
  viability of, 23
Sporidesmium, 73
Sporocarps, blue-green algal, 95
Spruce sawfly, 7
Steinernema, 7
Sterile compost, 73
  for mycorrhiza, 55
Strain selection, 6
  in blue-green algae, 93
Streptomyces, 8
Stylo, 37, 40
Stylosanthes, 41
Sugar beet, 16
Sulphide injury of rice, 90
Super-producers of antibiotics, 103
Suppressive soil, 72
Survival,
  of inoculum, 107
  of rhizobia, 36
  of antagonists, 69
Symbionts, mutualistic, 49

Symbiotic associations of blue-green algae, 95
Symptomless infection by virus, 79
Systemic aquired resistance to virus, 83
Systemic infection of virus, 79, 84

Take all, 69, 72
Tea, 57
Teknar, 19
Temperate forestry, mycorrhiza and, 61
Temperature, *Azolla* growth and, 95
Thuricide, 5, 19
Thuringiensin toxin, 104 (*see also* Bt toxin)
Ti plasmid, 69, 72, 102
Tilapia in rice culture, 91
Tillage,
  biocontrol and, 67
  disease and, 4
  mycorrhiza and, 55
Tillering in rice, 92
Tn5 transposon mutants, 71
Tobacco,
  protoplast, blue-green algae and, 96
  transgenic plants in, 82
Tobacco callus, blue-green algae and, 96
Tobacco mosaic virus, 79
Tobacco rattle virus, 82
Tolerance to pesticides, 93
Tolypothrix, 90
  tenuis, 91
Tomato, 80
  transgenic plants in, 82
Tomato mosaic virus, 80
Tomato moth, 5
Toxic metals (*see also* Metals)
  mycorrhiza and, 53
  uptake of, 54
Toxicology, 4, 14, 20
Toxin, 7, 17
Toxin genes, 101
  expression of, 103
  from *Bacillus thuringiensis*, 105
Tra minus strain of *Agrobacterium*, 103
Trace elements,
  copper, 60
  deficiency of, 60
  uptake of, 3
Trade secrets, 24
Transfer deficient mutants of *Agrobacterium*, 103
Transgenic plants, 1, 82
Tree stems, 73
Treflan, 13

117

Index

Trefoil, 33–34, 36
Trialeurodes vaporariorum, 7
Trichoderma, 67–74
  harzianum, 73
  viride, 5
Tropical forestry, mycorrhiza and, 61
Tulipa, 84

Undersowing crops, 74
Uptake of nutrients, 49, 101

VA, *see* Vesicular arbuscular mycorrhiza
Vacant niche, 72, 74
  for colonization, 68
Vaccinium, 51
  corymbosum, 59
Variability,
  of inocula, 21
  of results, 74
  of results in field, 58, 71
Varietal resistance to disease, 68
Vectobac, 19
Vertalec, 5, 19, 103
Verticillium lecanii, 5, 7, 19, 103
Vesicular arbuscular mycorrhiza, 50
  gene manipulation in, 101
  market size for, 61
Vetch, 31–35
Viability,
  commercial, 17, 20
  of control agents, 23
  of CP & E agents, 20
  of inoculum in production, 21, 103–104
Viable count, 21
Vietnam, *Azolla* culture in, 94
Vigna unguiculata, 44
Vine, 73
Virus, 3, 79–87
  coat protein of, 81–82
  control of pests by, 7
  diseases due to, 3
  future of inoculants, 85
  genetic engineering and, 24, 84
  inoculation with, 79, 80, 84
  localized infection by, 83
  mild strains of, 80
  systemic infection by, 83
  vectors of, 79

Walnut, 5
Wax moth, 6
Water availability,
  Azolla and, 95
  blue-green algae and, 95
Water potential, control agents and, 74

Water relations, mycorrhiza and, 53, 58
Weeds,
  control of, 3
    by microorganisms, 14, 106
  in rice culture, 90
Wheat, 16, 57
Whitefly, 5, 7
Wild-type organisms, patents and, 24
Wilt disease, 72
Winged bean, 31–32, 39–40
Wollea, 90
World, use of *Rhizobium* in, 29
Wound inoculation by *Agrobacterium*, 103
Wound sites,
  for biocontrol, 68, 73

Xenorhabdus, 7

Yam, 32
Yield,
  effect of *Rhizobium* on, 30–31, 39
  factors limiting in legumes, 39, 41
  of crops, 11, 20–22
  of rice, 91–92
  virus inoculation and, 80

Zinc, 53–54, 60
Zineb, 93
Zygomycetes, 51